PIONEERING SPACE

PIONEERING SPACE

— SPACE —

LIVING ON THE NEXT FRONTIER

James E. Oberg and
Alcestis R. Oberg

Foreword by Isaac Asimov

McGRAW-HILL BOOK COMPANY

New York   St. Louis   San Francisco
Toronto   Hamburg   Mexico

1 2 3 4 5 6 7 8 9 DOC DOC 8 7 6 5

ISBN 0-07-048034-6

Library of Congress Cataloging in Publication Data

Oberg, James E., 1944-
Pioneering space.
Includes Index.
1. Space stations   I. Oberg, Alcestis R.,
1949-   II. Title
TL797.033   1986   919.9   85-16587
ISBN 0-07-048034-6

Book design by Kathryn Parise

The opinions expressed in this book are solely those of James and Alcestis Oberg and cannot be construed to reflect any official positions of NASA, McDonnell Douglas, or any other agency, corporation, or association.

The extensive quotations from Valeriy Ryumin's diaries are by kind permission of Mr. Henry Gris, sole U.S. agent for this copyrighted material.

Permission from Mrs. Virginia Heinlein is gratefully acknowledged for quotation from "The Notebooks of Lazarus Long" in Time Enough For Love by Robert Heinlein.

TO OUR ANCESTORS,
WHO BROUGHT US TO THIS THRESHOLD,
AND TO OUR DESCENDANTS,
WHO WILL CROSS OVER IT.

contents

—Contents—

foreword

ADVANCE
— into the —
KNOWN

Isaac Asimov

L ife—*all* life—has tended to be adventurous, whether consciously or not.

About 650 million years ago, living cells, having remained single and separate for over two billion years, launched the experiment of multicellularity. About 400 million years ago, sea creatures began venturing out onto dry land, which, till then, had been sterile. Perhaps 10 or 20 million years ago, certain primates ventured out of the trees and became ground-dwellers.

In every case, of course, intelligence was limited or, by our own standards, nonexistent. What was done was done without possible forethought, and was done over a long period of time. It was done with who knows what losses, what fiascos, what harvests of death, and came who knows how close to overall failure.

Then came modern man, and he was adventurous, too, breaking new ground and extending the human range. Some

25,000 years ago, human beings ventured out of Asia eastward into the empty (of humanity) American continents and south-eastward into equally empty Australia. In early historic times, Phoenician ships ventured out of the Mediterranean into the Atlantic ocean. Later, Polynesian vessels crisscrossed the vast Pacific in magnificent feats of island-hopping. Still later, Vikings plied the unknown northern seas.

In these cases, it was intelligent beings that were involved, people who could weigh what they were doing—but in no case did they (or *could* they) know what awaited them. The early Siberians did not know what unexpected dangers might exist in North America. The early Polynesians did not even know for certain that they might actually come across an island. They survived, however (at who knows what cost), and these human explorations took place far more quickly than earlier nonhuman victories had taken place.

In the last great wave of exploration, between 1400 and 1900, when Europeans explored all the coastlines and continental interiors of the world, leaving only Greenland and Antarctica for the twentieth century, matters went even more rapidly, yet there was still no certainty as to what might be found. The explorers might hope they would be finding wealth in the form of gold or trade—and sometimes they did—but they might also fear they would be finding hostile and perhaps powerful natives—and sometimes they found that, too.

Now, as the twentieth century winds to its close, we stand at the brink of another great wave of exploration, another stride into a new and greater realm; and it is something that dwarfs all that has gone before.

All earlier experiments of life—all expansions of range—have at least been confined to Earth. There remained certain factors of environment that remained constant: there was always the ocean, always the atmosphere, always the Sun and the Earth's rotation, always the seasons and buoyancy and gravitational pull. Life had always held on to these things.

But now, as we launch ourselves into the greatest adventure

of all, we are abandoning Earth for infinite space. We are going beyond the air, beyond the water, into strange realms of the abnormal. We can't rely on a surrounding ocean of air, or water always within reach, or the comforting temperatures that are neither too high nor too low. Even gravitation itself fails us.

Yet despite this, we go with a rational confidence that no preceding explorers, human or otherwise, can possibly have had. There will be losses, we can be sure, but they will be fewer losses and on a smaller scale than exploration has ever seen, and we can be sure of that.

Why this confidence? Are we perhaps being lured into over-weening and dangerous overconfidence?

No, we are right to be confident, for we have what no earlier explorers had; not Captain Cook; not Columbus; not Leif Ericsson; not Hanno the Phoenician; and certainly not that first small living thing that lifted itself onto land and with-stood the down-beating of undiluted sunlight. We have knowledge.

We know the land we are invading. We understand the laws that govern it, and have known them since the days of Newton three hundred years ago. We know the lands that lie beyond, thanks to the instruments we have manufactured—from telescopes to rocket probes.

We even know some of the problems that will face us in day-to-day life in space for human beings have already traveled to the Moon, and have remained in space for up to eight months at a time. What's more, these explorations have been the common adventure of the United States and the Soviet Union, the nations which, in all other respects, are firm and resolute antagonists. (So perhaps the cooperation and understanding that elude us on Earth may be reached in the otherwise unfriendly emptiness of space.)

Here, in this book, the Obergs, who have been painstaking observers of both American and Soviet space exploration through all its as-yet short history, summarize what we know about life in space and what we may expect as such life multiplies and expands.

They are both comprehensive and comprehensible. They talk soberly and interestingly of the great imponderables such as light, and warmth, and air, and food—and the nitty-gritties such as toilets and privacy. The physical factors may be solved, but what of the psychological factors? That gets full treatment.

No one reading this book can fail to be impressed by the difficulties that lie before us; or fail to be even more greatly impressed by the excitement and opportunities. Surely, we cannot fail to see that stepping across this new threshold (as the Obergs phrase it) is the most exciting thing that will have happened to humanity so far, and the most worthwhile.

acknowledgements

Any endeavor of this scope involves not only our work but the efforts and understanding of people who touch our lives, both personally and professionally. We wish to thank the people who helped us through all this:

Our families—especially Jean Oberg, Nicholas and Demetra Ritsos, and Sandy Graff—who cheerfully postponed Thanksgiving and Christmas 1984 until April and May 1985.

Our son, Gregory James, for whom one traditional Christmas went down a black hole, along with the backyard fort we promised to build him.

Our infant son, John Nicholas, who had to give up breast-feeding for the sake of this book.

Dr. Dan Woodard, M.D., whose friendship and gallows humor we've long enjoyed, and whose guidance on medical aspects of spaceflight is accurate as it is thought-provoking.

Nicholas Timacheff, whose expert perspective and painstakingly precise commentary was of great and unique value to the preparation of this book.

Albert Harrison, for his invaluable help on the psychological aspects of spaceflight, and especially in taking time to read the text and make crucial and sometimes hilarious comments on it. His work on *Living Aloft*, along with that of his colleagues Mary Connors and Faren Akins, will be a fountainhead of ideas from which generations of space psychologists will draw inspiration.

—Acknowledgements —

Mark Lardas and Nelson Thompson, for their painstaking and constructively critical review of the manuscript.

Georg von Tiesenhausen, with whom we've had several very exciting and mind-expanding conversations. His help with source material and perspectives on robotics were essential to the robotics chapter.

Jack Paris, who patiently guided a novice through the ins and outs of remote sensing, and with help on source material we would have missed entirely without his help. May NASA fund all his excellent and worthwhile projects forever.

Mary Connors, whose conversation opened new vistas on the effects of spaceflight on human communication and life in general.

Joe Rowe and Ed Zvetina, for providing Soviet space source material far above and beyond the call of official duties.

Mike Gentry and Lisa Vazquez, for space photographs in Houston, plus the public relations specialists at McDonnell-Douglas, Grumman, Boeing, Lockheed, Motorola, and—last but not least—the Information Office of the Soviet Embassy in Washington, D.C.

Larry Bell, Jack Stuster, and Henry Fuhrmann, for providing us with valuable insights and documents.

Henry Gris, for access to and permission for excerpting from the copyrighted Ryumin space diaries.

Richard Curtis, our excellent and long-suffering agent, who patiently went over every word, syllable, and nuance of our contract.

Leslie Meredith, our fine editor at McGraw-Hill, who went to bat for this book. A thousand thanks.

Cynthia Merman, our copyeditor, who touched the manuscript only to improve it.

To our marriage, which proved that a couple can work synergistically, and that the totality of our effort—and life— together is far greater than the sum total of our work—and existence—separately. We both grew while working on this book, but not in one another's shadow.

introduction

Aboard the permanent manned-orbital outposts of the near future, there will be new ways of working, of seeing, of breathing, of thinking. In the contrast between new forms of ordinary human activities and the ways things have been done before, a new threshold in thought will be crossed, and a new arena of human activity will be embraced.

So far, voyages into space have been characterized by their temporary nature. There have been quick dashes to the Moon, not unlike the races to the North and South poles early in this century. There have been marathon orbital endurance runs, glorified space campouts with well-defined endings. These accomplishments have not lacked courage, ingenuity, and productive results. What they have lacked is permanence.

The first steps in this direction have already been taken. America's Skylab space platform was our world's first successful venture in long-term space endurance; in 1973–74, a series of teams of astronauts conducted productive activities aboard the module for as long as 84 days. The Soviets also demonstrated the depth and breadth of their commitment by launching a sequence of small Salyut space stations for occupancy by cosmonauts. Participation broadened in 1983 when the first Spacelab modules were carried into orbit aboard space shuttle missions; although brief, these missions were characterized by a vast amount of advanced equipment, participation of true scientists and specialists (who were not professional astronauts), and contributions of the European Space Agency that built the module.

These activities will continue throughout the 1980s, leading to the as yet uncrossed threshold of a permanent human presence in space. After that moment there will never again be a time when all Earthborn life is restricted to a single world in the universe.

The arrival of this stage—and the public perception that this stage has arrived, which is bound to follow soon afterward—will have a fundamental impact on our concepts of our world, our universe, and ourselves. Practical benefits will accrue to the nations directly involved; philosophical insights will be available to humanity as a whole.

The precise technology and architecture of the tools and habitats of space stations are not particularly crucial to the main benefits such activities will bring. Numerous designs— a "Space Operations Center," a "Power Tower," a "space train," a "honeycomb alignment"—have been drawn up, and no doubt many will be built. But the important thing, which this book stresses, is what can and will be experienced aboard these facilities, what will be the essential features of human life in orbit.

The volume of Russian material on space-flight psychology is astounding, as is their cosmonauts' candor in discussing many delicate aspects of human behavior in orbit. Long excerpts from inflight diaries—by Lebedev, Aleksandrov, Ryumin, Kubasov, Savinykh, and others—have been published in the U.S.S.R., along with in-depth interviews and postflight reports of other cosmonauts' impressions. Few Americans on Skylab kept diaries, and none has ever been published.

Besides, between 1971 and 1984 there were fewer than six-hundred man-days of American space station experience, compared to almost four thousand Soviet man-days aboard a succession of Salyut vehicles. Consequently, much of this book is based on Soviet material. But the kinds of experiences and insights described are human, not national or ideological or even parochially culture-specific. In the 1990s, more than a thousand human beings from more than a score of nations will be sojourning in space on months-long missions, and they will add a symphony of experiences to these sketchy outlines.

While it cannot be overlooked that these diaries were written by a subset of Soviet spacefarers—only the civilian flight engineers (not the pilots), and only those on the scientific (not the military) Salyut missions—still and all the material is unique in the annals of human exploration, invaluable in preparation for American space station efforts of the early 1990s, and quite assuredly authentic (if noticeably incomplete).

Because of the general unfamiliarity of the spacefarers' names and their mission details, the appendices include brief biographies and mission descriptions of the people and events referred to in this book. And we use "Soviet" to refer to government functions, and "Russian" to refer to ethnic subjects.

As the population base for American spacefarers widens, more impressionable and more expressive individuals will experience spaceflight and will candidly reveal their feelings and insights. That is a natural process, and it is already occurring, but an awareness of the vast sweep of the Russian spaceflight soul barings is bound to accelerate it. Both trends contribute to a better public appreciation of the spaceflight experience, and a more rapid impact on our entire culture.

The space experience is an essentially human one, deeply rooted in our history; the technological trappings are merely secondary, however much easier they are to display to the public. Space is being pioneered by our souls as well as by our machines. That is the theme of this book.

chapter one

SPACEFARERS' DIARY

When is the exact moment one crosses the threshold between Earth and space? Perhaps it is when one first perceives the huge bending of the horizon, the first glimpse of the Earth's curvature. Or perhaps it is when one lets go of a pencil and it floats, as American astronaut William Lenoir and Russian cosmonaut Valeriy Ryumin did. The first perception of crossing over is whenever the mind perceives the departure from the familiar. "There was convincing evidence of weightlessness as soon as I released the flight plan or a pencil," Ryumin noted in his inflight diary. For the first time, the mind verifies the bizarre, alien, unearthliness of space, and says, "Yes, this is it. I am here, in outer space."

Once the cosmonauts had docked their spacecraft to the station, they prepared to go over into it. The transfer tunnel was filled with air, and one of them entered it to unlatch the station's front hatch, which had spent the previous several months exposed to airless space and searing sunlight. Later he recalled a momentary impression: "In the docking unit we sensed the odor of burnt steel—the odor of space."

Cosmonauts Lyakhov and Ryumin woke at eight in the morning to the sound of a "disgusting siren." After exercising for half an hour, they washed up, and had a breakfast of canned meat, cottage cheese from a tube, a roll, and instant coffee. The coffee is made from their own recycled perspiration.

After three missions in which he (and everyone else) had to defecate into plastic bags, Pete Conrad was particularly excited about the zero-g toilet aboard Skylab. He rushed to "christen" it, taking a stack of checklists with him. An hour and a half later he emerged, triumphant. "It worked great!" he announced. "And I'm now the first man to circumnavigate the whole planet riding on a toilet seat!" The astronauts did notice a design flaw: the reading light was on the ceiling, but the hole was on the wall, requiring a posture that placed one's back to the light, casting a shadow onto the reading material. Later, Conrad also rode a bicycle exerciser for another ninety minutes, all the way around the world.

The space shuttle pirouettes and somersaults with slow grace throughout its flight like a magnificent performing animal. The pilot can direct its small-attitude rockets to poise itself silently and gracefully—nose straight down toward the ground, tail down, wingtip down, belly down, top down.

The spacefarers sometimes glance out the window, and seem startled to realize that the Earth is not where they expected it to be. A momentary disorientation grips their guts. There is no up and down, unless a person says to himself or herself, "Up is that which is above me, and down is that which is below me." The universe is no longer heliocentric to the mind, or geocentric to the senses. In this universe of no bearings, the reference frame has to become anthropocentric; all spatial references can make sense only with reference to one's body. In this disoriented world, an astronaut might reach out for the security of old reference points. Joe Allen looked down at Earth sites, and said to himself, "If this is east, that must be north." But "up," "down," "east," and "north" are meaningless. They are the vocabulary of yesterday.

Veteran cosmonaut and mission control center flight director Aleksey Yeliseyev called veteran Valeriy Ryumin out of the blue, asking him to take Lebedev's place as the flight engineer of a seven-month spaceflight. Lebedev had torn his knee ligaments while working out on a trampoline, and required surgery. It was only four weeks until blastoff, and the backup crew was unready. "Think it over, and tomorrow we'll talk about it," said Yeliseyev to the stunned Ryumin.

"Did I want to fly? Could I fly through that time period with a new commander? How would the news be received at home?" Ryumin asked himself. The hardest question to answer was the last. "My wife, Natasha, and my mother simply began to cry and their attitude seemed as if I voluntarily condemned myself to execution. The kids, too, were adamantly against it. All their summer vacation plans were instantly destroyed."

Fear of the unexpected and unforeseen was always near at hand. Aleksandrov and Lyakhov were alarmed by a loud crack, like the sound of a shotgun. "One of those micrometeorites paid us a visit right in the middle of a communications session with the Earth," recalled Aleksandrov. "We heard someone knocking at the door!" he joked. It had smashed into a window, and left its signature—a dark spot—across its surface.

One cosmonaut of a two-man space team sneaked off in the middle of the afternoon for a nap, picking the powered-down Soyuz capsule for his resting place. Later, his shipmate began looking for him. Noting the closed hatch, he politely knocked. The first man woke, startled, and groggily responded, "Who's there?" Both spent the next several minutes in uncontrollable giggles.

Valentin Lebedev and Anatoliy Berezovoy slept little the night before their spacewalk. The next morning, after meticulous spacesuit checks, Lebedev opened the hatch: "It was like a light snow on a frosty day; dust from the station was floating around the station in the form of tiny sparklets. . . . Along with the dust the tiny washers and nuts that had somehow got lost flew out of the casing of the compartment; a pencil floated by.

"My first feeling on opening the hatch was a huge Earth and a real sense of the unreality of everything that was happening. All around there is perfect silence, no sense of the speed of the flight. No wind whistles in your ears, nothing weighs on you. The silence is striking. The station is frozen in space like a block. At 0620 we went into shade. The station hatch shone like an open door in a house in the countryside. All around was the dense blackness of the space night."

A universe devoid of substance and light is frightening. "Speaking frankly," said Ryumin, "I was not exactly eager to step out into the void. It was scary. I had to 'stand up' in front of the opening prior to exiting, and although this makes no difference in weightlessness, psychologically it was a little disturbing." He clung to the Salyut, motionless, as darkness fell. When the sun rose a half hour later, he moved slowly toward a troublesome antenna, "wrapped in thought about the impending difficulties facing us." His crewmate played out the lifeline, the umbilical, which looked like a large intestine against the background of Earth.

Jack Lousma's memory of spacewalking: "To be hanging by your feet as you plunge into darkness, when you can't see your hands in front of your face—you see nothing but flashing thunderstorms and stars—that's one of the minutes I'd like to recapture and remember forever."

Zeitgebers, the physical, physiological, and social measures of time, are torn to pieces in space. The rhythm of life is no longer measured by fundamental perceptions of light and darkness, nor are the seasons perceived tactilely as hot or cold. "Sometimes there are no nights here," said cosmonaut Aleksandr Aleksandrov. "The sun shines all the time through the ports. It is very difficult to get used to the situation in which, in a short period of several orbits, it is possible to see the change in the time of the year and the day and night periods, dust storms and snow drifts. . . . In an hour and a half, we travel right around the world, and time is compressed so that the minutes can cover a year."

Fears of the possible but unlikely also lurk within. "I was afraid most of all of an appendicitis attack," said Ryumin. "I still had my appendix. Also, I was afraid of getting a toothache. Once I dreamed I had a toothache. I awoke almost instantly, feeling, yes, my tooth really did hurt. But, by morning, the pain was gone."

To maintain enough strength to return safely to Earth, long-term spacefarers must exercise vigorously. Even the most avid physical fitness fanatics were discouraged in space. "How joyful the morning warm-up was on Earth," recalled 211-day veteran Berezovoy. "But here the sweat really runs. And it turns out that exercise on the bicycle or treadmill is not a pleasant form of rest, but exhausting labor on which a great deal of working time must be spent." A year later, 150-day spacefarer Aleksandrov noted: "You get warm quickly, the face and arms sweaty, and by the end of the training you are soaked. But the legs are dry. Then we sponge down with a warm, damp towel. The perspiration is not salty, but rather bitter. But you are thirsty like under Earth conditions." Skylab astronauts recalled the disgusting way a pool of sweat half an inch deep would accumulate over their breastbones and just slosh around there, with nowhere to run "down" to.

The astronauts got into a good-natured banter about their impressions of viewing Earth. "There sure is a lot of open water!" remarked Air Force Major Mike Mullane. "Mullane's gained a lot of respect for the Navy," joked naval Lieutenant Commander Mike Coats. "Well," drawled Hank Hartsfield, the mission commander, an ex–Air Force officer himself, "there's air over it all. . . ."

"We could always count on something funny happening to us. For instance, my wife had sent me a box of chocolates [in the supply ship], her favorite kind. Me, I can't stand chocolate. I threw the chocolates behind a panel. Before getting into the station again, we stocked it up and brought in clean linen. All of a sudden the box of chocolates jumped out of nowhere. We didn't know what to do with it. Kovalyonok inadvertently opened it, and the chocolates flew all over the station like peas. We spent two hours trying to retrieve them. However, the next crew—Ryumin and Lyakhov—found two of the chocolates. As it turned out, they had had a brandy filling. . . ."—Aleksandr Ivanchenkov.

In the space station, all the sounds they hear, apart from their own voices, are mechanical: the whirring of fans, the hum of computers, and the inhalation of uptake vents. Anatoliy Berezovoy loved the cassette recordings of natural sounds that seemed to give his space life a missing dimension: "We had recordings of sounds: thunder, rain, the singing of birds. We switched them on most frequently of all, and we never grew tired of them. They were like meetings with Earth."

Transcript of a space-to-Earth conversation:
  "Can Joseph say 'Daddy'?"
  "Da-dee!"
  "Can Joseph say 'Mommy'?"
  "Ma-mee!"
  "Can Joseph say 'one'?"
  "Wun!"
  "What does a cow say?"
  "Moo-oo!"
  "What does a dog say?"
  "Bow-Wow!"

The best part of spaceflight? "Why, sleeping!" say Lyakhov and Aleksandrov, "especially when there are dreams." One dreams of being with his son, fishing, catching a gigantic, prize fish. Another dreams of skiing, and awakens to a cool breeze from a fan blowing on his face.

"About day 45 we had what you might call the first sensitivity session in space," Gerald Carr recalled after his 84-day mission in 1973–74. Other observers called it a "space strike." The men, frazzled by a stream of exhausting orders from Mission Control, had decided to take a day off to get their heads together and give Earth a chance to calm down. Carr was a great believer in separation as a psychological balm: "When people feel like a wart, they have to go somewhere they can be a wart," he theorized.

While Leonid Kizim orbited far above the Earth, his wife bore the daughter conceived late the previous summer, six months before his flight. "Congratulations," said flight director Ryumin, "you are the proud father of a baby daughter." They showed him a picture of the infant over the cabin television, but it would be half a year more before Kizim would be able to hold her in his arms and wrap her tiny hand around his chubby index finger.

"That film vault, it's a god-awful mess. I'm sitting here looking at it now, and it just kinda makes me wince. Every time I have to go over there to do something, I cringe and I say, 'Oh, God, I gotta go through this again.' . . .You open up the cover, and then you got .cameras and lenses and film and tapes coming out, and it's just like the top of a snake pit, and they all start slithering out at you. . . ."—Ed Gibson

Astute observations from space save lives. In August 1983, Radio Moscow reported that the then-orbiting Salyut-7 team had warned of flood danger in the Pamir Mountains. From space, they "were able to observe the formation of a big mountain lake created by the intensive melting of glaciers. The water threatened to flood several mountain hamlets. A timely warning from orbit made it possible to evacuate the inhabitants from the threatened areas and a channel was promptly dug to empty the lake gradually." In many other parts of those mountains, however, melting of the glaciers caused by a heat-wave led to rivers bursting their banks and flooding roads.

After several weeks in space, the crewmembers' spines tended to straighten and lengthen. Each man on Skylab gained an inch or more of height, which caused problems getting into their snug spacesuits. But 5′6″ Pete Conrad didn't care: "At last," he exulted, "I'm taller than my wife!"

"One of the videocassettes we were sent showed my daughter's birthday celebration. It also included some familiar Moscow scenes. We couldn't wait to get that videotape. I would put it on when I felt particularly homesick. You watch, get engrossed in it, and it seems like you're with your family."— Anatoliy Berezovoy

At the big wardroom window, the Skylab astronauts unconsciously adopted a geocentric posture: whichever way the space station was pointed, they would rotate around the rim of the window until they were level with Earth's horizon. This often meant they had to sit crossways in reference to the wardroom's "up-down," or even directly "upside down." That bothered them less than being cattywampus to Mother Earth herself.

Astronaut Ron McNair carefully assembled the saxophone he had stowed along with his personal goods. He began a fine, very skilled, very sensitive jazz solo. As he did, he floated free in the shuttle middeck, as disembodied as the notes he played.
    Years, continents, and cultures away, two cosmonauts had strummed a guitar sent up as a surprise in a spare corner of an unmanned supply ship. "We were momentarily overcome with nostalgia," one of them recalled. "But then our circumstances yanked us back to reality."

"I remember when Leonid Popov arrived as part of a visiting expedition," wrote Valentin Lebedev after his own 211-day expedition. "He shared some of his 185-day flight experience with us and said there were times when he was overcome with depression. We had already been up three months at that time and had not experienced any such critical situation. I don't even think we had any afterwards."

During reentry into the Earth's atmosphere, a large reddish-orange apparition appeared outside the shuttle's overhead windows. "It was an incredible standing shock wave, shaped like those monoliths on Easter Island," recalled astronaut Jeff Hoffman. "Its head was somewhat mushroom-shaped, and right in the middle, where it necks down, was a very intense point of white radiating light. It was like a hallucinogenic vision, just spectacular. I called Rhea [astronaut Rhea Seddon, his crewmate] over to see it. She looked out, and her eyes opened wide. We looked at each other and she said, 'Do you think we should bow down and pray to it?'" It loomed above them as the shuttle plunged toward Earth, radiating its alien glow steadily, even after they passed from darkness into sunlight.

When Lyakhov and Ryumin blasted off, the sky was a sullen gray. A heavy snow had fallen only the day before. "I looked around at the snowy, gloomy wintry sky. I wanted to absorb everything, to remember it, to preserve it for the entire flight," said Ryumin.

After landing, the men were mildly surprised to see Earth in the ripeness of oncoming autumn, almost harvest time, as if they had unconsciously expected the Earth to remain true, like an old lover, to their last precious vision of it. Even though from space they could plainly see the changes of the seasons, the changes were a reality apart, a reality they could not share in until they returned. They knew they were back when, as Ryumin said, "I sensed Earth's gravity sitting heavily on my shoulders." Someone placed simple daisies in their arms. "How pleasant it was to hold them," said Ryumin, "to breathe in their fragrance."

The seven astronauts were headed back to earth from their Spacelab-3 expedition. In the front seats, the commander and his co-pilot read off the spaceship's atmospheric entry and descent milestones for the benefit of the five scientists who did not have display screens in view.

The deceleration forces squashed the men down into their seats. After the effortless agility of weightlessness—the fulfillment of the universal human dream of flying—the returning astronauts felt increasingly oppressed by the gradually mounting sensation of weight on their arms, legs, chest, neck, throughout their bodies. Just when the crushing weight seemed to mount to an unbearable strain, the veteran spacefarer commander called out cheerfully: "Men, we're now at one-G. You are going to feel this way for the rest of your lives!"

Don Lind, one of the scientists, vividly recalls the lance of agony this thought shot through his soul. "This is how the guy must feel when he wakes up from anesthesia and discovers they've amputated his legs," he agonized to himself. "I'm going to feel crippled the rest of my life!" A fleeting vision of the soaring freedom of human motion in space tantalized him and then faded with the growing weight on his chest. To return from space was to resume a style of motion he had happily but all too briefly learned to forget. The freedom of spacefaring was gone.

"Modern exploration had to be an adventure of the mind, a thrust of someone's imagination, before it became a worldwide adventure of seafaring. The great modern adventure—exploring—first had to be undertaken in the brain. The pioneer explorer was one lonely man thinking."—Daniel J. Boorstin, *The Discoverers*

chapter two
COMMUNICATIONS

Contact Light! Okay, engine stop. Aye-see-aye out of detent. Mode control both auto. Descent engine command override off. Engine arm off. Four-thirteen is 'IN.' . . .—*Buzz Aldrin, 1969, mankind's first words from the Moon*

Houston, Tranquility Base: the Eagle has landed.—*Neil Armstrong, 1969, mankind's second words from the Moon*

To listen to a conversation between space travelers is to be faced with a foreign language. At Mission Control in Houston, one person alone is allowed to talk with the "space people." This person is designated as CAPCOM, a word that formerly was merely an abbreviation for "capsule communicator." But now, as with so many others in the vocabulary of astronauts, the word has taken on a life of its own in "spacefarer pidgin."

Our everyday experience is that social kinship can be disconnected from appearance, from skin color, even from mode of dress—but not from speech. We can immediately recognize strangers by an accent or verbal performance. According to such criteria, today's space travelers are already "strangers" to their parent society.

Certainly, the rise of jargon is not unique to spaceflight. Every technical profession has it, and many subcultures within a national culture utilize—even cultivate—distinctive modes of speech and vocabulary. There is a special terminology or slang among pilots, musicians, numismatists, teenagers, Marxists, ethnic communities, geographic regions. These dialects serve to economize on specialized communications, avoiding ambiguities that in the language of the whole society would require extensive circumlocutions to resolve. But jargons are also often deliberately cultivated by subcultures to enforce exclusivity. The fact that conversation is unintelligible to outsiders may not be an accident at all.

So far, no such alienation seems to lie behind the special language of spaceflight. Literally far-out gadgets and concepts require bizarre-sounding terminology. But the potential is there. And whenever the words are different, a difference in thought processes follows. If an Eskimo has a hundred words for "snow," and only one for "tree," a dictionary may be insufficient to discuss even such everyday subjects as the weather or lumberjacking.

Even more conducive to misunderstanding is a situation where the words remain the same in two cultures, but their meanings have diverged. It's hard enough for an American to go to England and discuss boots, bonnets, valves, napkins, and other shared terms with distinctly different meanings. A similar problem is arising with "spacefarer pidgin," where words like "abort," "destruct," "assembly," and "depress" mean something very different from their "English" equivalents.

In terms of everyday English, most astronauts seem stiff, unimaginative, and word-poor. The best they seem to say when confronted with literally unearthly vistas is, "What a beautiful view." But in terms of the language that ensures their survival, the language of spaceship functions and navigation and "failure modes," fluency and even poetry thrive.

Seeing this phenomenon as the consequence of a diverging language—and inevitably of diverging thought patterns—may help introduce an important theme. The space experience is *different* from every previous earthside experience (however many strained analogies may be conjured up), and this experience will affect human consciousness in wide-ranging and surprising ways. When enough people see Earth and its relationship to the universe differently, major new ways of thinking about things—about Earth, about humankind, about intelligence, about time and space themselves—will appear.

The changes in communications styles brought about by the space environment and the spaceflight experience do not just make communication between space crews and earthside operators more difficult. Hitherto unrecognized problems in communication between crewmembers are also introduced.

It's not only words that can change their meanings in space. Among what Soviet cosmonaut Valeriy Ryumin (veteran of two six-month space voyages) called the "charms of weightlessness" are severe distortions imposed on human communication in space. In weightlessness the face swells so that, "seen in a mirror, [our faces] were difficult to recognize." But on Earth, facial expression is a fundamental aspect of communication, and visual cues are the first weathervanes of human mood we learn to read as babies. In space, the face remains swollen throughout one's stay, so this fundamental nonverbal communication technique is nullified.

Not only does puffiness obscure emotive facial expression, but aboard spacecraft under weightlessness faces are seen sideways or upside down, in odd and unfamiliar angles and unusual lighting conditions. Subtle facial signals become indistinguishable, their often subliminal messages lost. Spacefarers say that this can lead to misunderstandings and the proverbial "failure to communicate."

Russian doctors V. I. Myasnikov, E. F. Panchenkova, and F. N. Uskov have tried to study facial distortion because "puffiness may affect interpersonal intercourse in the same way as primary disturbances, such as shifts in the emotional-volitional sphere." Their methods of study included measuring the contours of the face, and setting up a means to take such measurements quickly in flight using a grid projected over the television image of the face. After some study, they developed a technique to determine which soft-facial tissues play which communications roles when various emotions are being felt.

Using television cameras aboard the Salyut space station, these scientists noted both the spatial distance between the cosmonauts and their eye contact. Both factors play an important part in nonverbal facial communication, and help define the general atmosphere of the group and the mutual emotional attitudes.

Nonverbal communication in space was virgin territory for the psychologists. Nobody seemed to have paid it much attention during Skylab (1973–74). The Russians in the late 1970s and early 1980s were careful not to formulate hasty conclusions from their observations. Much effort went into perfecting the experiment itself since they had found that placement of cameras and the plane in which the image was seen were both associated with erroneous evaluations by ground-based psychologists.

The other great nonverbal communication technique is "body language," assuming special postures or relative positions. This too is essentially impossible under spaceflight conditions. One still can make oneself understood with exaggerated miming gestures, but posture, proximity between speakers, small gestures, and subtle body movements—which on Earth are all important communicators of mood, distance, and closure—are rendered impossible in space. This is also virgin soil for the psychologist, an area in which spacefarers will develop their own responses and means of communication, not yet apparent and certainly not yet even universally understood among today's astronauts and cosmonauts.

Pure verbal communication is distorted in unexpected and unprecedented ways. Because of the fluid shift to the spacefarer's head, voice pitch is changed and becomes more nasal. The rhythms of dialogue can also change. Again, all of these speech factors are on Earth important indicators of our mental state. They are verbal cues we are all taught to recognize unconsciously from earliest childhood.

Because of these space-induced variations to earthside normal voice patterns, changes in pitch and tone alone turned out to be imperfect indicators of mental stress in space. Until two-way television communication was set up on Salyut-6 in 1979, Russian psychologists mistook elevated speech expressiveness as a symptom of increased emotional "reactivity." Two-way television allowed ground personnel to communicate better with space crews, with less area for misinterpretation, and specialists recognized these factors as normal adaptations, not symptoms of emotional stress ("normal" only in a spacefarer frame of reference, of course).

Nonetheless, even with two-way television and the improved understanding of the part voice pitch plays in space, there was still room for errors. People on the ground involuntarily associated altered pitch and tone with heightened tension. In the memoirs of his 211-day flight, Anatoliy Berezovoy pointed out: "The proper tone must be set in conversations with the ground and in your relations with your comrades. . . . Sometimes not everything in the conversations is understood and they hear only the tone, and they get the wrong impression: the cosmonauts are irritated, giving trouble. In fact, of course, nothing of the sort is happening. The conversation had to do with business, and here, other intonations are possible."

When international crews were added to Soviet and American spaceflights, verbal communication was further complicated. For one, the foreign visitors had to learn the spacefarer jargon (Russian spacefarer pidgin is as arcane as American). These languages of acronyms and code names are not easily mastered even by native speakers, and they certainly can be difficult to understand in the short training period foreign visitors are apt to undergo. The Czech visitor aboard Salyut-6 in 1978, Vladimir Remek, commented that even when languages are as close as Russian and Czech, there is still a certain amount of "interior translation" involved. This, he recalled, "more or less slows down the communication process." He went on to point out that even if visitors speak the host language fluently, they may have "articulation habits that lead to a deformation of expression, thus making the communication more difficult."

Most important is the way in which language guides the formation of thoughts. "Together with the mother tongue," Remek explained, "every human being obtains a certain way of thinking, perceiving the world of things. These particular features form his mental ties with the remaining members of the same linguistic community. In certain conditions it might be reflected in behavior differences requiring a mutual under-standing in order to prevent a disruption of the harmony among the crew members."

Remek found that after fifteen months of training, he was unable to translate Russian spacefarer pidgin into his mother tongue when he broadcast a speech to his own people. He noted that his speech was characterized by occasional omis-sions, repeated words, words disturbed by stammering, slips of the tongue, "the hesitation phenomenon" (about a third of his communication during his television appearance was an "er" or "a"), and a rise in pitch.

Considering this experience, the Czech guest cosmonaut advised standardizing space terms to a greater degree, and, during training, giving a visitor adequate terminology in his or her mother tongue to communicate with fellow citizens. He also recommended selecting linguistically adept spacefarers, and taking into account linguistic and cultural differences.

It has never been easy for the tightly knit astronaut or cos-monaut groups to fully accept culturally different foreign vis-itors, no matter what goodwill may lie behind the concept of international flights. During training for the joint Soviet-French flight in 1982, one cosmonaut was withdrawn from the crew because of his refusal to treat a non-Soviet as a full-fledged, equal partner in the crew's activities; but at the same time, Soviet space officials set up curtains in the cosmonaut caf-eteria to isolate the French crew from even the sight of other people. During training for the Soviet–East German flight, the Soviets were somewhat astonished and intimidated by the professional intensity of their East German visitors. When two West European Spacelab astronauts trained in Houston as space shuttle "mission specialists" in 1980–82, they could infer from the behavior of the hierarchy that they were con-sidered outsiders, however warm were personal relation-ships.

Seven-month spacefarers Berezovoy and Lebedev pointed out in 1982 that they found themselves more at ease with their own countrymen, no matter how well an international orbital visit had gone. "We all felt somewhat constrained [with the international crew, including Jean-Loup Chretien from France]. But when Lesha Popov, Svetlana Savitskaya and Sasha Serebrov [their second visiting crew] arrived, everything was quite gay and simple, as if it were our own brothers and sister who had dropped in to see us. There was no end to the jokes and laughs."

This is not to say that the Russians are inordinately provincial; the same type of social dynamic occurs on American flights. "In any group," said Mary Connors, a communications expert at NASA-Ames, "even someone slightly different will tend to be isolated from a closely knit group." Spaceflight crews tend to be "mature groups," people who have known one another for many years, have worked in the same circumstances, and have very similar viewpoints and language. When an individual from outside the group is thrown in, for a period of a few months during training, he or she is never completely integrated into the group.

"I will never be part of the astronaut club," noted Charles Walker, the first commercial engineer to fly with a shuttle crew in mid-1984. Nor did he try to "join the club" by crashing the invisible barriers. Characteristically, during the inflight onboard television transmissions, while the off-duty professional astronauts cavorted and played, Walker remained literally in the background, perpetually tweaking the somewhat troublesome pharmaceutical-manufacturing equipment he had come aboard to oversee and service. And after the flight, the astronauts expressed gratitude that Walker had not acted like a tourist or gotten in the way during any of their three satellite launches. He had abided by the dynamics of the group, and his own professionalism had earned him respect and affection—but not intimacy.

Just how exclusive the brotherhood of astronauts can be was made visually clear recently. As a rule, crew patches are made up of a center medallion with a symbolic picture of some sort and a black circular border with the names of all the crewmembers. For Spacelab-1 in late 1983, and STS-41D in mid-1984, the names of the "outsiders"—scientists Byron Lichtenberg and Ulf Merbold in the former and Charles Walker in the latter—were embroidered on the bordering circle along with those of the regular NASA astronauts. But starting in late 1984, the names of those outside the astronaut club (for example, Canadian Marc Garneau and oceanographer Paul Scully-Power on STS-41G, Air Force Major Gary Payton on STS-51C, Senator Jake Garn on STS-51D, and French "spationaute" Patrick Baudry on STS-51G) were embroidered on a separate banner attached below the circle of astronauts' names. This occurred even though there usually was plenty of room for inclusion in the literal "inner circle." If a picture is worth a thousand words, this graphic message might be that it's a rough road ahead for any outsider to presume to full-fledged inclusion in this group of space professionals.

Nothing is certain more quickly to bring on condescending amusement and contempt from a group than for a stranger to attempt to ingratiate himself through the use of "exclusive" jargon. As soon as parents and teachers and television commercials begin adopting teenage slang, teenagers begin to abandon the terminology for newer expressions. Among many military units in war, "newbies" do not have the right to call their comrades by their first names, or to be so called themselves, until they have been "baptized by fire," or undergone a combat situation. At NASA press conferences in Houston, participants visibly wince when newscasters who have not earned the respect of the astronauts aspire to camaraderie by a forced—and usually laughably inaccurate—recourse to spacefarer pidgin. Such attempts to break down the barriers of jargon are doomed to failure because the intruders have utterly failed to understand the psychological value of such jargon: it helps keep them at a safe distance until they have shown worthiness to approach the brotherhood.

In space, the feeling of being an exclusive entity is further reinforced in those occasional "us" versus "them" skirmishes that both Russian and American space crews have experienced with their respective Mission Controls. Cosmonaut Vitaliy Sevastyanov, who made a two-month trip in 1975 before retiring to host a popular Soviet television show on spaceflight, complained of the "perpetual nitpicking," and astronaut Robert Parker (Spacelab-1, 1983) snapped angrily at scientists on the ground for altering and confusing established flight-experiment procedures. The famous "Skylab-4 Strike" in 1973 has been overblown, but the actual event was in fact symptomatic of Earth-space tensions and festering ill will.

Part of the problem is the medium in which these spacefarers were communicating: direct "mediated" communication. In other words, they used somewhat public, radioed speech. Such a communication system has turned out to be adequate for giving and receiving information, and for businesslike formal exchanges. And in fact daily business on Earth is usually carried out under mediated circumstances (that is, over the phone) in which negotiating stances are somewhat preordained, clearly planned and articulated, and somewhat rigid.

"The audio-only medium may have an advantage of perceived privacy," and has the advantage of being high on substance and low on showmanship, concluded Mary Connors and space psychologists Albert Harrison and Faren Akins, authors of an important 1985 space-psychology study entitled *Living Aloft*. Computer/telegraphic communications are even more formal modes of exchange, although, like all written forms of communication, they can result in the most intransigent positions. However, because these two forms tend to emphasize rational rather than social considerations, they are preferred by people who wish to filter out irrelevant and irrational interpersonal "noise." The result, said Connors, citing media expert Robert Johansen, is "enhancing the communication of highly informed 'pure reason'—a quest of philosophers since ancient times."

In space, most business communication from the ground

is carried over teleprinter or computer text/graphics (the most intransigent of forms), and by audio-only communications (the second most formal). These media lack the ability to convey visually the frantic pace at which spacefarers are usually trying to complete the already established required activities. While adequate under most circumstances, these media exacerbate problems that do arise. A certain radio etiquette has to be maintained in this medium or there can be misunderstanding and hostility.

And that's not all. According to Connors, spaceflight conditions distort fundamental communications in many ways. In space (unlike on Earth), business communications between crewmembers are carried on face-to-face, which is a medium of compromise and potential socioemotional content where information is carried across visual, auditory, and tactile channels of communications. However, intimate communication with family members is forced into the more public, formal, and distant medium of audio only. This is likely to have profound implications for future space communications schemes.

The Russians sought to ameliorate this fundamental limitation of audio-only communication medium with the addition of a two-way visual system. At first, they had a system in which the mission controllers could see the cosmonauts but the cosmonauts couldn't see the mission controllers (this method currently exists also on the American space shuttle, as it did on Apollo and Skylab). As the *Living Aloft* study pointed out, such a system has problems. The crew can feel as if they are being surveilled, and unannounced eavesdropping may be seen as an unwarranted invasion of privacy. The Americans sidestepped this issue by letting the astronauts turn the TV system on and off; they also can select which of many cameras are used, so no "outsiders" can cause conflicts over privacy and timing.

In 1979, the Soviets installed a two-way television system aboard their two-man space station. With it, the cosmonauts could see people on the ground, and vice versa, in real time. This is as close to face-to-face communication as anybody can get, although camera angles, size of screen, and quality of picture interfere to an extent with communication. It was a great benefit, of course, for the cosmonauts to be able to speak in this manner to their families, a practice that they performed with unflagging regularity throughout the subsequent long-term flights.

Although these family gatherings were regarded by all as an overall psychological boon, they had their drawbacks. "Tears flowed freely," wrote *Pravda* correspondent Vladimir Gubarev, after listening in to a session. Nevertheless, the cosmonauts looked on these highly charged moments as "cathartic," chances to unload their emotional baggage. And, afterward, there was sometimes a sense of depression and letdown because the spacefarers knew it would be a week before the opportunity to talk would come around again.

As the flights lengthened in duration, so also did the communications sessions. The families had been coached not to unload any bad news. Ryumin noted in the diary of his first long-duration flight that conversations with family members would not be absolutely candid or truthful: "Our questions were, naturally: 'How are things at home? How are the kids doing in school?' Even though we knew that our wives would certainly not give us anything but good news. We could not help them anyway."

And Ryumin was the person who would be most aware of this. Early in Georgiy Grechko's 1977 flight, his father died. Ryumin had been in Mission Control in Moscow, and must have been aware of the decision not to inform Grechko directly. When cosmonaut Vladimir Dzhanibekov visited the space station a few weeks later, he quietly informed crewmember Romanenko of the death. Romanenko did not tell Grechko the news until their return to Earth months later.

Psychologists both here and in the Soviet Union are questioning whether future space communications should be facilitated (by various means such as more frequent and even voluntary calls home) or inhibited. Those who argue against close family ties being maintained during long-term space-flights claim that emotionally charged family communications may threaten the cohesiveness of the space crew, and that worriers may neglect their duties and instead dwell on earth-side issues. Those who argue for closer ties to families claim that although space crews generally get along together, they still maintain a certain personal distance. "The Skylab crews did not become as brothers," pointed out NASA historian W. David Compton, and there has been nothing to suggest that most Russian crews found any intimacy in space beyond the needs of the profession. These people would argue that certain fundamental family relationships are necessary to physical and mental well-being, and maintaining those intimate relationships, however inadequately at long distances, will keep the individual sounder of mind and body than if he or she were cut off from family, or if family communication were carefully censored and sanitized.

The paradox of space communications is that, despite the cohesiveness and exclusivity the long-term spacefarers maintain among themselves against all outsiders, they still seem to maintain an emotional distance among themselves, and they are careful not to cross over an invisible threshold between professionalism and intimacy. They look to Earth with an odd mixture of longing and alienation.

Should a spacefarer require psychological counseling, there are a number of options open to him. Crews might spontaneously create what Connors, Harrison, and Akins referred to as "a buddy system." Each crewmember might have a special individual in which he or she could confide. Counseling could even be done remotely over private phone or T.V. link-up. Both media have advantages: some people find it easier to confide in a disembodied voice than a person, and others like face-to-face interviews that they can control, more or less, by electronic means.

As spacefarers venture farther from Earth, these bonds must be maintained, even nurtured, against mind-numbing distances across which even lightspeed signals trek for many minutes. Voyagers to Mars, for example, would face radio signal round-trip times of between five and twenty-five minutes.

Under such conditions, even simple conversations with earthside associates and loved ones are impossible. Yet for effective communications, and for the spacefarer's mental health, something better than alternating monologues is called for.

One technological trick could satisfy many of the human needs for conversation, for seeing how people at the other end of the line react to each other's comments. Here's how it might work:

One side starts off a message, then goes off and does some other task for ten or twenty minutes. An alert tone annunciates the incoming reply, which begins with a replay of that side's original voiced comments against a visual image of the faces of the people at the other end, at the exact moment they were hearing the first side's words. They then launch into their comments, giving the first side the impression of immediate response, that is, of conversation. The process is repeated from his or her point of view, and over several hours—interspersed with other productive activities—a complete conversation is built up. This conversation, mimicking a face-to-face no-time-lag process, can then be played back by the far-voyaging spacefarer, in public or in private as he or she prefers.

It would be a strange way to "converse," a sort of two-way trick time machine. But space communications will have seen even stranger variations of the way, until now, humans on Earth have "always done things." Off Earth, spacefarers will do this, and dozens of other basic human functions, very, very differently.

chapter three

— The —
SEASONS
— of —
SPACE

It's been an interesting morning to wake up. It was beautiful, sunny, and we pulled up the window shade and looked out there. And about halfway through the pass, the sun went down and now it's black as night. I guess we'll roll over and go back to bed.—*Jack Lousma, 1982, aboard the space shuttle*

The white nights have started in orbit. The Sun barely goes below the horizon. And its edge slowly slides behind the orange cover of the horizon like a curtain in a puppet theatre.—*Valentin Lebedev, 1982, aboard Salyut-7*

When humans lived exclusively on Earth's surface, they used sunrise-sunset to mark their days. They watched phases of the Moon to mark "months," and divided the flow into four even parts marked by easily distinguishable events, eventually called "weeks." As the daylight periods lengthened and shortened, they marked their seasons. Later, they invented schemes and mechanisms to mimic these natural patterns, and called them calendars and clocks.

Spacefarers have brought these timekeepers and rhythm-markers with them, having almost forgotten their original connection with natural events. In space, new rhythms are appearing. But the verdict is not yet in on the way these rhythms are accommodated, poorly or well as the case may be, by our anachronistic timekeeping methods.

For a space station orbiting Earth at low altitude—say, three hundred miles high, with an orbital plane inclined to Earth's equator by twenty-eight degrees—there still are sunrises and sunsets, but they occur so frequently, every ninety minutes or so, that they cannot serve as days in any traditional sense. During twenty-four hours a space station inhabitant would experience about sixteen such "days."

Just as on Earth, some activities are constrained to "day" or "night," and some are not. Sleeping, for example, is divorced from conditions of external illumination, and on small space vehicles great care has to be taken to install lightproof sunshades on windows lest the "noon" Sun wake a sleeper.

But seeing through the windows does depend on "day" or "night"—daytime for seeing targets on Earth's surface below, or for watching the Sun itself; and nighttime for observing other astronomical targets, such as stars, planets, or the aurora. And solar-powered vehicles depend on the day/night cycle first to gather power (some for immediate use and some to charge batteries), and then to switch to the batteries as darkness approaches—or, more accurately, as they approach the darkness.

On Earth, knowing the time of day gives a good clue about the sunlight conditions, but this is not true in space. Spacefarers must carry day/night schedules with them in order to plan their daylight- and darkness-dependent activities to correspond to "clock time." Orbital day and night are completely divorced from earthside (and human, based on five billion years of hereditary experience) "day" and "night."

One solution to this would be to start each 90-minute "space hour" at sunrise, or, as with "days" on Earth, at midnight. A small problem is that as the satellite's altitude shifts, its period of revolution changes by several seconds. And, as on Earth, one would have to know the season to know how much of a day was "day" and how much was "night." On Earth, the two phases average out over the course of a year to be the same length; in orbit, because a satellite is still sunlit long after the ground below is dark (due to its altitude), on average there is about 60 percent "day" and 40 percent "night"—but as with the "midnight sun" of the far north, there are orbital seasons in space when the Sun never sets.

It is the space station's orbit around Earth that creates these "space seasons." But they run through a full cycle—a "year," if you will—in only a few weeks. Furthermore, another cycle, which coincides in length with the earthside "year," increases the severity of these orbital seasons.

This requires some explanation. Because of Earth's equatorial bulge (it is slightly squashed out of round), the plane in which a satellite orbits Earth is slowly twisted. Viewed from an Earth-centered frame of reference, this motion is toward the west by several degrees per day.

To visualize this, perform an experiment. Place a hardcover book on a table, and spin it around as if it were nailed to the table with a spike through its center. Lift the cover a bit—the cover's angle with the table is analogous to the satellite's orbital inclination around the planet—and now as the book spins about that imaginary spike, the cover's "twisting" motion is the same as that which a satellite's orbital plane goes through (called "precession").

In practice, if a satellite crosses the equator at, say, 120 degrees west longitude on one day, and it completes exactly sixteen ninety-minute orbits in the next twenty-four hours, it will surprisingly not recross the equator at that same spot. It will instead cross the equator at somewhere west of that point, perhaps 125 or 130 degrees west longitude. This is a direct result of the precession caused by gravitational twisting.

The practical effect of this twisting is that a satellite retraces its ground-referenced path earlier (as measured by clocks) every day. Thus after a number of weeks the satellite's schedule has crept through a twenty-four-hour Earth cycle, and is back at its original position vis-à-vis Earth.

This is a very important cycle for satellites, which as yet has no name. For the Russian Salyut space station, it takes about fifty-eight days; for the planned American space station, at a different inclination and altitude, it will be about forty-eight days.

For convenience, supply missions to future space stations will probably blast off at the same local launch-site clock time. This will standardize training, and also allow the use of standard mission flight plans. Such launchings, then, will be "seasonal," occurring during a regularly scheduled interval of several days every two months or so. Many parameters of the space station's pulse of life—crewmember duty cycles, water supplies, menu rotation, trash stowage, data collection—will tie directly to this seasonal rhythm.

If these supply launchings must occur every so many days, then even the physical design of the supply ships will reflect the rhythms: they will be built to carry n-man-days' worth of consumable supplies. Thus by a subtle but firmly linked chain of causation Earth's equatorial bulge determines how many changes of spacefarer underwear will be wrapped in each standard resupply pack!

Not all the rhythms will change, of course. Lunar cycles will remain important because of possible interference with astronomical observations or because of potential aid in low-light surface observations. And a spacefarer's twenty-four-hour "day" is not likely to change much, if at all.

Some science-fiction writers of a generation ago imagined that weightless sleep would be so soothing that only a few hours per "night" would be sufficient. This has not proven to be the case: eight or nine hours is still the norm in space, as on Earth.

Further subtracting from useful worktime are the requirements for meals (up to two hours a day) and exercise (another two hours). Several more hours are usually spent in communications with Earth, either live or in review of uplinked written instructions. That leaves a working day of about eight hours, just like at home (except that "leisure time" usually comes out on the short end of the clock!).

Some of the other patterns that have governed human activity for ten thousand years or more will still play themselves out in space, but on different time scales. For example, there will be "summer" and "winter" in low Earth orbit (LEO for short). And every "year" (fifty-eight or forty-eight or whatever days long) will have a "harvest" (supply ship visit) and a consequent "Thanksgiving dinner."

It's easy to envisage the orbit of a space station as it circles Earth. At one point it emerges from Earth's shadow, traverses the sunlit side, then plunges into the shadow again. Out of a ninety-minute circuit, fifty minutes or more are in sunlight and a bit less than forty are in darkness. But it can vary, and under some conditions, a satellite can spend its entire orbit in sunlight.

Remember the spiked book analogy. Lift the front cover about halfway—that represents the orbital plane, inclined to Earth's equator. Imagine now a circle drawn on the book cover. This represents the satellite's orbit, and the satellite swings around every hour and a half or so. How much of that circle is in sunlight and how much in darkness?

Let a light at the other end of your room represent the Sun. When the spine of the book is pointing more or less toward the light (when the Earth-Sun line is about in the plane of the orbit), the largest possible amount of the orbit is in darkness.

But rotate the book halfway about that imaginary spike, until the spine is crossways to the light. Also, imagine Earth tucked inside the circle, without much room to spare (the satellite's orbit is quite close to the Earth's surface). When viewed from the light, the orbit will appear to skim the rim of the Earth, dipping behind it for only a brief portion of each full circuit. This allows the smallest possible fraction of the satellite's orbit (or none of it) to pass through darkness.

To experience full sunlight, two conditions are necessary:

Earth itself must be tilted at a maximum toward or away from the Sun (that is, near the summer or winter solstices, June 21 or December 21), and the satellite's orbital inclination must be fairly high. For the satellite's orbital plane to be exactly face on to the incoming sunbeams, an inclination of at least fifty-seven degrees is needed, so that added to the twenty-three degrees of Earth's axial tilt gives you ninety degrees, a right angle.

In practice satellites in orbits with inclinations above fifty degrees can experience full orbital sunlight for a few days near midsummer and midwinter, since at their altitudes the sunlight "spills over" the edge of Earth and illuminates space objects that are some distance "down sun" from the sunset line on Earth's surface. Satellites with lower inclinations experience less pronounced "white nights" (analogous to midsummer short nights in Canada or northern Europe), but the rhythm still exists.

Aboard the Salyut-7 space station, Soviet cosmonauts have watched and described this phenomenon. Lebedev remarked that the Sun barely went below the horizon. On a spacewalk in November 1983, Aleksandrov noted how ". . . we saw how the glow, like on Earth during the Leningrad white nights up in the north, practically never disappears in one place and it is already appearing on the other side, so that practically the whole of the horizon is constantly illuminated."

In space life, these cycles work out this way: in every precession cycle, a satellite goes through a period of shorter-than-average orbital nights. The higher the orbital inclination, the shorter the night portion of the orbit. And twice an Earth year, around June and December, these short-night phases can become so extreme that for several days there is no night at all as viewed from the spacecraft.

Since we know that some activities are favored by daylight (like observation, or solar-panel power generation), it follows that such activities should preferentially occur at these short-night phases. And indeed they do, as exemplified by Soviet experience aboard the Salyut-7 space station in 1983–84.

When the time came for spacewalking cosmonauts to plug in additional solar panels to their habitat, for safety's sake they had to disconnect the wires of the panel (one of three) being augmented. For several hours, the other two panels had to supply the station's full power load. Under normal day/night cycles, this would not be possible, since at least a third of the panels' output has to be used to charge batteries for the next dark period. However, when the dark fraction dropped from a normal of forty to less than ten minutes, battery charging became so much less of a drain that two panels were adequate, and the third panel could be taken off line for maintenance.

These same light/dark cycle variations will be important for space gardens. (See Chapter 8). At some intervals in the precession cycle, the farm modules will receive considerably more sunlight than at others. The growing cycle of the plants will have to be properly phased to take advantage of this natural orbital sunlight rhythm.

With plants, and with people, it's not nice to mess around with Mother Nature's rhythms in orbit. Experience on Salyut missions in 1971–77 taught Soviet space doctors how not to plan crew sleep schedules.

Recall that the precession of a satellite's orbit brings it farther and farther west every twenty-four hours. While it may at first cross the Soviet Union, and be in radio contact for up to half an hour, that "high-visibility" pass occurs earlier every day. Other orbits pass south of the U.S.S.R., and cross few if any tracking sites; these are at first scheduled to occur while the crew is asleep, but they gradually also shift around the clock.

To ensure that more communications periods occurred while the cosmonauts were awake, Soviet space planners tried to be very clever: they simply instructed the spacefarers to wake up about half an hour earlier every day to synchronize their sleep cycles with the shifting communications phases. In practice, this forced the cosmonauts into a 23½-hour "day," which space doctors considered a minor, feasible adjustment.

But the spacefarers hated it. They awoke fatigued and irritable, and often had to take medication to fall asleep earlier every night. Their circadian rhythms were completely knocked out of whack.

As chief Soviet spacecraft engineer Konstantin Feoktistov admitted in 1981, these "sliding days" for the cosmonauts, based on "orbital time," led to "a negative effect on their well-being."

By 1976, when a space crew actually had to be evacuated prematurely for fatigue and stress, the "sliding day" plan was thoroughly discredited. Beginning aboard Salyut-6 in late 1977, the spacefarers experienced strict twenty-four-hour days, synchronized with Moscow Time (the same clock used by their Mission Control Center), and just accepted the occasional midday communications lapses.

Here the first venture to adapt earthborn organisms to alien space rhythms was a failure. Back on Earth, specialists in a new field called chronobiology should have been able to warn the Soviets: according to Dr. Charles Czeisler of Harvard University, the body's internal clock, if it doesn't get any external reset cues, tends to run a bit longer than twenty-four hours, not shorter. A worker who changes shifts can adapt more easily to later hours than to earlier ones, Czeisler found, and certainly the experience aboard Salyut-4 in 1975 bears this out.

One wonders what similar circadian troubles lurk far out in space. On Mars, for example, the "day" is twenty-four hours and forty minutes long. Flight controllers for the Viking robots in the late 1970s invented a new name for "Mars day"— "sol," from the Latin for Sun.

The Viking landers had no trouble adjusting to this slightly lengthened sunrise-to-sunrise period. But how about human beings? Will they be able to tweak their circadian rhythms sufficiently to sleep soundly and wake refreshed under sols instead of days? If Czeisler's theory about chronobiology is correct, there should be no problem. But between a million generations of ancestors, and a new world with a large potential of descendants, the issue of human rhythms becomes of cosmic significance.

Whatever their schedules, human beings must sleep. When a spacecraft is in flight continuously, this can present problems.

On the first "long" American manned spaceflight, Gemini-4 in 1965 (four whole days!), the two astronauts took turns sleeping in the cramped cabin. But so close were the quarters and so disturbing even the gentlest maneuvers and quietest radio conversations, that neither slept well. On all subsequent Gemini and Apollo missions, the entire crew slept simultaneously. As far as is known, the Soviets followed this same practice on their Soyuz missions. While the crews slept, flight controllers at the earthside Mission Control (in Houston, Moscow, and the Crimea) kept watch over the spacecraft.

On the long-term Skylab missions of 1973–74, the policy was continued: the crewmembers were awake together and asleep together. The operational convenience and psychological value were so persuasive that shift operations were never considered. And on Soviet Salyut space stations throughout the 1970s and early 1980s, the same approach was used.

The short space shuttle missions also adapted this traditional scheduling, since nighttime activities tended to wake sleeping crewmembers. But when the Spacelab module was introduced in late 1983, a new potential—and a new requirement—for shift activities appeared. First, the equipment was too expensive to leave idle half the time; second, the shuttle cabin could carry more crewmembers than could be kept busy at the same time; third, the laboratory section was physically isolated from the shuttle cabin, so crewmembers could sleep undisturbed even while others were busily (and noisily) at work. So a set of bunks was installed on the middeck of the shuttle, and two shifts—the "Red Team" and the "Blue Team" on Spacelab-1, or the "Gold Team" and the "Silver Team" on Spacelab-3, or any other choice of color codes—were organized.

It was the first space shift work in almost twenty years, and it worked very well. The off-duty crewmembers reported sleeping very soundly (occasional disturbances usually caused only temporary waking). While one team worked, the other had a few hours to review that day's activities and prepare for their next day's activities. Twice a day, the teams met for "handover," as results, problems, and changed checklists were discussed.

In practice aboard Spacelab missions, the personnel of the two teams share bunks (the "hot bunk" scheme of the submarine service—some military units even put three men sequentially into the same bunk every twenty-four hours!), except for the commander. He has a reserved bunk full-time, and "floats" between both shifts whenever crucial events are scheduled. For normal operations, other astronauts (usually the copilot and the flight engineer) are nominally "in command" of routine operations on each shift.

For brief periods, the Spacelab-1 spacefarers in December 1983 had experienced a new concept in orbit: they were really "off duty." Somebody else was fully responsible for running the spacecraft, so they could temporarily "stand down" from the tense "on duty" feeling of full and unrelenting responsibility borne by all their predecessors. Their spare time, such as it was, was their own. They could gaze out the window; one even brought along a small "ham" radio and talked briefly with earthside amateur operators (everybody from the kid next door to King Hussein of Jordan). Theirs was an unusually novel point of view in space, but one to be shared by more and more spacefarers in coming years.

Another ancient human rhythm is associated with days of the week. Every so often (every quarter lunar cycle or week, more or less), we are accustomed to a "day off." In the past, some explorers and colonists (such as in French Quebec) tried to increase workers' output by reducing the rest day to one in ten or even less frequently, but productivity went down. It is acceptable to "float" the rest day (say, Mondays instead of Sundays) on an occasional or even regular basis, but there must be some break in the work sequence, or efficiency suffers.

Space station crews generally get Sundays off, when they can turn their attention to some of their own tasks rather than follow orders from Earth. Early during the Salyut-6 expedition sequence, Soviet space doctors realized that a single day was insufficient, since it tended to get overloaded with tasks not done during the workweek. So a two-day weekend of "active rest" was instituted, along with special events such as televised family conversations. Long-duration Russian spacefarers have found this cycle quite adequate.

Upon all these cycles and seasons, the human rhythm must be superimposed. As with any major effort, a person must get used to it before he "hits his stride." And as with any intensive activity, there is a point beyond which fatigue and psychological stress begin to detract from efficiency.

There are distinct phases for a human voyaging in space. To operate most efficiently, spacefarers must literally learn to move and see all over again. "Our cosmonauts say that during their first month in orbit they are only learning to 'see' and to work productively," noted three-time space veteran Aleksey Yeliseyev in 1981, "because many things escape their notice at first."

The Byelorussian space pilot Vladimir Kovalyonok, veteran of a five-month and a three-month expedition, gave a more explicit breakdown of the stages in a 1979 interview. "The length of the flight is not just the qualitative growth of knowledge. The whole expedition may be divided into several periods. In the first month, one becomes accustomed to flight conditions and adjusts to them. The second month, one gradually begins to gain knowledge and collect facts. In the third month, one begins to analyze them more deeply and many questions arise. One begins to feel like a real researcher in the full sense of the word." It should be pointed out that even as he discussed the new rhythms of spaceflight, Kovalyonok still instinctively utilized earthside measures of time.

This phenomenon of increasing adaptation to working in space was addressed by leading spacecraft designer and former cosmonaut Konstantin Feoktistov: "The efficiency of prolonged orbital flights increases, on the whole, two-fold or three-fold as compared with short space flights."

After about a month, the spacefarer is at or near maximum efficiency. But after three or four months, psychological and physiological fatigue sets in, and efficiency drops. "The very long flights," Feoktistov later admitted, "were purchased at the price of declining crew efficiency." These "Space Ages of Man" remain largely uncharted, but at least their existence has been noticed.

Even the spacefarer's perception of the flow of time and the relation of the present to the past and the future can seem altered. "Little by little the rhythm of life is settling into a daily routine" noted cosmonaut Aleksandr Aleksandrov in his flight diary for July 6, 1983, the twelfth day of his 150-day expedition. "But there is something that separates it from the usual, earthside way. Everything is remembered as if it was very long ago and we are not together and it is not known when it will be again."

This sense of a separate cycle from Earth persists and grows as time in space passes. After three months, not yet at even the halfway mark of his seven-month space voyage in 1982, Valentin Lebedev wrote: "Only in two months will I be able to even think about or hope for our landing. . . . Will I ever really be back on Earth among my family, and will everything really be fine again?"

But alongside all of these new and difficult patterns, some syncronization with Earth remained. First, of course, the twenty-four-hour day. And viewing Earth reminded the men of their homeworld's natural seasons. "Visual observations from space were for us a unique way of communing with our own nature, which brought exactly the same kind of joy as on Earth," recalled Berezovoy. "From orbit we observed all the seasons of the year; the launch was in the spring, we flew throughout the summer and fall, and the start of winter. It was very interesting to trace the way in which our planet moved through the procession of the seasons. At first the whiteness gave way to the onrush of the greenness, and then gold covered the fields and forests, and then the whiteness again, only now its frontier was moving in the opposite direction."

The spacefarers saw Earth's natural seasonal variations, but were outside of them, and so felt them somehow "unreal." Returning to Earth after many months, cosmonauts consistently testify to the "calendar shock" of seasonal weather at the landing site on the steppes of Soviet Central Asia. To leave in the spring, and return to driving snow, left vivid impressions of momentary disequilibrium until the spacefarers realigned their internal concept of the proper season with Earth's actual situation.

For the course of their spacefaring, the cosmonauts had been outside Earth's rhythms, wrapped in their own. Spacefarers are as yet only dimly aware of these alien days, months, seasons, and years—but as expeditions lengthen, the calendar shock on returnees will grow.

Until, that is, the point when a spacefarer on Earth only briefly between voyages will automatically and unconsciously maintain his or her "body clock" on space rhythms, and consider earthside patterns of time merely a temporary interlude. That person will be the first human being tuned in to the universe and not merely a local planet—and what patterns these new rhythms take will be for him or her to discover.

chapter four

EARTHWATCHER'S FIELD GUIDE

When you look out, you see things in three dimensions. You would think that from orbit the globe would look rather flat, as though you were in a high-flying airplane. But, in fact, that is not the case. You see a third dimension very clearly. You can see the mountains coming up from the surface of the Earth. You can see the clouds well above the surface of the Earth. You can see the shadow of the clouds on the Earth. That shadow, that relief of clouds and mountains and Earth below is very striking, and gives you a sharp feeling of three dimensions. That third dimension is really an overwhelming aspect of the view from orbit—*Joe Allen, shuttle astronaut, 1983.*

Baykal was a particularly delightful sight. The lakes of Africa are also pretty. South America has two jasper-colored lakes in the 40-degree latitude area. That was interesting, certainly, but there's nothing better than Baykal. It's different every time—but always pretty. It is important to understand in general that for us in orbit, visual observation replaced theatre and movies. It provided the opportunity to take a breather from our other work and relate one-on-one with the Earth.—*Anatoliy Berezovoy, Salyut-7 cosmonaut, 1982*

For the three astronauts of Skylab in late 1974, the observations were baffling but challenging. Something odd was going on over the Indian Ocean, something nobody had ever noticed before.

The first evidence was the unusual clouds. They were unlike any clouds the men had seen anywhere else in the skies of Earth, whether viewed from below or, from space, from above. In density and shape they were similar to cirrus formations, but they appeared much darker than usual. The astronauts could not determine how high the clouds were in the atmosphere. Two or three orbital passes brought the space station over the area, so the men observed the cloud formations, which stretched for thousands of miles, at different times of day and from different angles. They got the gradual impression that the clouds looked "smokey."

Thousands of miles to the west, over Africa, the astronauts also were seeing dark smudges in the daylight. But when their orbits took them across Africa at night, a different image appeared. Then they saw a bright strip, about four hundred miles long, running through Tanzania and Kenya. By day, the strip was obscured by dark, moving plume patterns. But closer eyeball examination showed that the strip ran near agricultural regions.

The spacefarers concluded—and later investigation verified this—that they were seeing slash-and-burn agriculture on a scale never before recognized. The hazy clouds thousands of miles downwind were smoke palls from those man-made fires.

Such widescale fires are not as obvious as we might expect. In 1983, while Russian cosmonauts spent almost half the year in orbit, the largest natural forest fire in history devastated Borneo, but there is no indication that the Soviets noticed it.

From such heights, only the gods of mythology contemplated Earth. For millennia, humans tried to imagine their viewpoint. What did they see? Now for the first time we have a good idea.

Colors and shapes, certainly, which came out on film as only comparatively vague shadows and subdued hues, if the camera caught the view at all. But this new view of the world was revealed most completely to the human eye. According to spacefarers' accounts, the three-dimensionality was equally striking. Their forward motion enhanced this perspective, with clouds over terrain of noticeable relief. At that altitude, one-mile relief at the surface is proportional to looking at the ground near your feet at half an inch's vertical difference.

Thus, with one eye shut (no true stereoscopic perspective, of course, at orbital altitude), and by moving your head horizontally at two inches a second, you would create a pseudostereoscopic impression similar to the one experienced in space.

On Skylab, one unexpected realization (aside from the fantastic acuity of the men's vision) was the individual difference in perceptual ability. "Each crewman saw sea-surface features that the others did not notice," recalled a scientist, but "each recognized the feature that intrigued the others when the texture or color was indicated." Gerald Carr saw "textile seas," while Ed Gibson saw "long, discontinuous waves" (actually windrows, or Langmuir cells). Bill Pogue discriminated subtle color distinctions more easily than the others. Their conferences at the observing window substantially increased the features recognized.

American astronauts have noticed that one particular viewing angle that earthside photographers deliberately avoid turns out to be a valuable feature in space: sunglint, or "glitter patterns." Skylab program scientists deemed it "an extremely productive method for observing the surface texture of the oceans and rivers of Earth."

Skylab astronauts saw many large-scale surface features of the oceans for the first time in this sunglint effect. The postflight science report gave details: "To take complete advantage of this phenomenon, the crewman had to recognize and identify the feature of interest in the sunglint pattern. The crewman had approximately five seconds to observe a feature in sunglint, to decide the appropriate observational technique, and to perform the technique for the unexpected feature. Although the occurrence of sunglint is predictable, the features that will be visible in the sunglint are not. . . . Certain features are more evident in the outer edges of the sunglint pattern than they are in the central area." The technique proved the best way to observe current boundaries, zones of turbulence, and slicks of all kinds.

Skylab astronauts noticed something that fifteen years of weather satellites had missed: the eyewalls of some tropical storms tilt outward. "This important result requires new thinking in tropical-storm studies," the scientists announced. "Computer models of tropical storms must include the possibility of sloping eyewalls; presently, they do not." Other eyeball descriptions of fine-scale eye structure were made as well.

Some of the spacefarer discoveries involved extremely subtle perceptions. For five months in 1978, Vladimir Koval-yonok observed regions of suspected geologic ring structures in the far southwestern U.S.S.R.. "The crew noticed seasonal variability," he later noted in his formal report in 1981, "in the color contrasts of plant growth and a characteristic distribution of morning fog formations. Observations made with different levels of illumination contributed to this success."

The oceans were particularly magnificent targets of observation. On Skylab, Gerald Carr tracked the ocean currents by sight. "I looked out to sea and got an absolutely perfect [view] of the confluence of the Falkland and the South Equatorial currents," he reported after a pass across South America (his shipmates had termed it a "surrealistic painting"). "You can see the Falklands Current coming up from the south, and you can see, not quite so clearly, the Equatorial Current coming down from the north. And right straight out from the city of Buenos Aires, you can see where these two currents meet and head straight out to the southeast. The colors down south are more of a powdery aqua, toward the chartreuse, whereas coming down from the north more of a deeper blue, turquoise or aqua. Where the two meet and head out, you have a mixture of two colors, and it is very streaky, very taffylike, and serpentine. The water itself is a good blue." "It was totally unexpected that the Falkland Current would appear bright green," marveled an environmental scientist afterward.

These kinds of observations help identify the new oceanic phenomenon of "cold water eddies," localized upwelling regions that affect the poleward transport of heat by the oceans, and consequently influence global weather and climate models.

Photographs of such phenomena (such as the Westward Drift current south of Capetown, South Africa) did not turn out. Pogue explained that "the tone sensitivity of the color film emulsion was inadequate." But the spacefarers' eyeballs worked fine!

Soviet impressions are similar: "No form of optics can replace the human eye," noted Vladimir Lyakhov, veteran of almost an entire *year* of orbiting the planet on two flights. "All the photographs and panoramic shots obtained from orbit are merely a pale shadow of what we were able to observe."

One Soviet conversation about the sight of the ocean from orbit went from the specific to the general. "The Sea of Okhotsk is brownish green, and the Caspian is a deep, dark blue. On our preceding pass we saw unusually colored spots in the ocean." Asked how the ocean looked currently, the spacefarer continued, "It is multicolored, iridescent, patterned, as if it were alive. We saw the track of a ship. Guadaloupe Island seemed to be swimming in the ocean!" In 1984, space shuttle astronauts independently specifically mentioned Guadaloupe Island as being trailed downwind by complex swirl cloud patterns.

Where the Gulf Stream and the Labrador Current meet off the coast of eastern Canada, the eddies and vortexes are also visible to the naked eye. This is because the oxygen-rich cold water has a light green-blue color, in contrast to the violet-blue warm water. Soviet oceanographers soon also noticed that many areas of the oceans had the most diverse hues, from greenish-yellow to reddish, which characterize the systems of major currents, the boundaries of the rising and sinking of waters, and suspended matter carried into the ocean by the largest rivers. Kovalyonok recalled that the yellow and brown river outflows distinguished themselves clearly against the ocean's blue-green surface. Pogue recalled how the Yangtze often turned the East China Sea brown for a hundred miles from shore; he also noticed that Amazon River tributaries each had a distinctive silt coloration (apparently associated with soil zones at the watersheds), and that these color distinctions were traceable for several hundred miles after a stream emptied into the main river complex.

Skylab astronauts in 1973 were the first to notice the green smudges on the oceans, caused by upwellings of plankton ("plankton blooms"). The ecologically devastating "red tides" are special types of plankton blooms, and the astronauts saw several such phenomena. The third crew noted evidence for a new branch of the New Zealand Current near the Chatham Islands, and this observation of hitherto unknown plankton blooms stimulated enough interest among New Zealand fishermen that surface exploration of the area began even before the astronauts had landed.

Half a world away, other American spacefarers had noticed the strange colors of the water around the Cape Verde Islands. Current-generated turbulence at the lee of each island was clearly visible. A dark area leeward of Santo Antao Island was realized to be a region of upwelling, which developed as surface water was moved away from the coast by the Canary Current and replaced by cold water from below. Cold water brings nutrients to the surface, and these areas are usually good fishing grounds.

Cosmonauts of Salyut-6, five years later, watched "fish slicks," oily films appearing in spots where schools of fish congregate. These in turn could be calibrated with water temperatures.

The importance to deep-sea fishing of human eyeballs in orbit was realized by the Soviets in 1978. The commander of that 140-day mission was Vladimir Kovalyonok, and in his mission report he described it this way: "During observation, areas were detected whose color differed markedly from that of the surrounding ocean. It turned out that such color differences were linked to the accumulation of plankton and were therefore of interest to the fishing industry. The observation method previously applied by the crew for ocean investigation was not used, and the results which were obtained turned out to be unexpected and did not agree with current concepts of ocean optics." The Soviet oceanographers were probably unaware of the Skylab observations five years earlier.

Kovalyonok reported seeing the borderline between cold and warm water near the Kurile Islands, northeast of Japan, and a rich yield of mackerel was later reported there. In 1983 at a postflight press conference, Lyakhov boasted: "We studied a number of regions of the world's oceans for signs of bioproductivity, and gave precise coordinate references for nine such regions. . . . In all nine regions, work by fish search vessels confirmed our information."

One wonder that should have been visible from orbit was not mentioned publicly: bioluminescence in seawater. This effect is caused by the disturbance of plankton microorganisms by the passage of pressure waves, perhaps due to passing ships or submarines. The appropriate plankton species live only in tropical seas, such as the Arabian and the Caribbean. The absence of any reference to bioluminescence has led some Pentagon oceanologists to fear that the Soviets may have found a military use for such observations, particularly in the detection of submerged submarines. According to a report in 1982, "The United States doesn't know how, or whether, this can be done."

Icebergs were often seen from orbit. From Skylab, the extent of icebergs in the South Atlantic was plotted better than ever before. In fact it turned out that it was easy to distinguish between old, disintegrating bergs, and sturdy, fresh ones that had not been long in the sea. Around the older, well-melted ones was a distinctively colored halo of fresh water.

On Skylab, the crewmembers had made sequential photographs and commentary of ice formation and movement in the Gulf of St. Lawrence, Hudson Bay, the sea north of Hokkaido, and the Great Lakes. The St. Lawrence studies revealed many ice characteristics that surprised those who had studied the area for years. One scientist who sponsored the study made daily aircraft flights during Skylab observation opportunities, and testified to the absolute superiority of the photographic and eyeball data; he predicted that regular use of space observations would enable limited navigation of the St. Lawrence throughout the winter.

To describe why the eye behaves in these ways, scientists analyzed the strengths and weaknesses of the human sensory system. In the mid-1970s, Skylab astronaut William Pogue developed a touring lecture program with the material entitled "The Eyes Have It." He may have been biased—but he'd been there.

Pogue first listed the limitations: "A scene is retained by the mind for approximately a sixteenth of a second," he reported. "This makes possible the illusion of motion by a motion picture projector [but] it also limits a human's ability to distinguish rapid changes in a scene."

Retinal sensitivity is also restricted to the narrow band of wavelengths that pass most easily through the atmosphere. In space, the sky is bright with flaring colors to which the eyeball is blind.

"At any given time," the astronaut continued, "one can concentrate on only one small area of a scene in view. Simultaneous events or different objects in the field of view become progressively more difficult to assimilate or correlate as their number increases."

Fourth, Pogue listed resolution: "The human eye can differentiate small, closely-spaced objects of a certain minimum size and spacing [but] after the resolution limit is reached, the eye sees a color tone average between the objects and background."

Last, while the eye can focus on objects various distances away, it cannot do so simultaneously, and must refocus to view a new object at a different distance within a few dozen feet (beyond that range, all objects are viewed "at infinity").

Each of these limitations can be overcome through mechanical aids, Pogue enumerated. Shutter, scanning, and strobe techniques can be used to stop motion down to a millionth of a second. Chemical emulsions and radiation-sensitive elements can record images at energy levels and wavelengths undetectable by the eye. A camera can record a scene indefinitely.

The advantages of a brain-eyeball perception system are amazing. The unappreciated features, judged Pogue, are "superior to any optical-mechanical-electronic apparatus."

The first feature is dynamic range. This is the widest difference in feature brightness that can coexist in a field of view without causing degraded perception of either extreme.

Associated with dynamic range is the ability to distinguish minute differences in hue and contrast. Photographs taken from Skylab, while stunning to ground personnel, were disappointing to the astronauts: "The reaction of all crewmen was the same," Pogue recalled. "The pictures looked 'washed out' and did not record feature details of texture and hue that were so clearly evident to the unaided eye. In some cases, items that were clearly visible to the eye didn't even show up in the picture, although proper exposure parameters were used." In particular, he noted that "pictures taken at low sun angles were a tremendous disappointment since the camera could not record what the eye interpreted as a reasonably good scene."

Pogue and other American spacefarers (oddly, the phenomenon does not appear in Soviet accounts) have described what has come to be called the "picket-fence effect," in which elements of cloud-obscured terrain can be assembled in the mind as the observer sees the screening features move across the face of the land. "The crewman's ability to mentally remove the effects of different conditions of haze, viewing angle, and illumination is unique," concluded the Skylab scientists. "In manmade geometric field patterns, colors, sizes, and shapes can be detected and classified through as much as 75-percent cloud cover." From space the typical subjects of such observation were geologic and geographic features, and weather phenomena. The brain's ability to perform pattern recognition, both of expected and of unexpected patterns, is a feature impossible to incorporate into automatic scanning devices.

Another surface property Pogue called "texture identification." Despite identical coloration of the observed targets, crew members were able to differentiate among adjacent areas of snow, ice, and clouds via textural distinctions—some sort of subtle pattern recognition beyond the capacity of automated sensors. His list of examples included sea and lake ice formations, sand dune structures, dust storms, air pollution, haze, and smoke palls—all distinctive features to the eye that could not be differentiated on photographs.

Speaking to the same kind of audience in the Soviet Union in 1981, Vladimir Kovalyonok drew on similar experiences, and made parallel conclusions. "High contrast sensitivity and color differentiation are characteristics of the eye," he reported. "Its resolving power reaches a single angular minute." The cosmonaut declared that while the eye has a spectral resolution of a few nanometers in wavelength, the best photographic equipment is twenty times coarser. The minimum amount of contrast necessary to differentiate terrestrial details is about 4 percent for cosmonaut eyes, but as much as 30 percent for equipment. Humans could reliably observe in conditions of solar illumination ten to a thousand times less than those needed by automatic systems, and at considerably lower angles of illumination (the utility of lower angles increased the total effective observing time in orbit of humans over instruments by about 30 percent, he claimed). The dynamic range that Pogue mentioned was, according to Soviet scientists, about ten orders of magnitude from dimmest to brightest.

Skylab scientists endorsed the view of the spacefarers themselves. Wrote oceanographer Robert E. Stevenson, whose protégé, Paul Scully-Power, went into space himself a decade later, "Only one of the major features observed by Skylab-4 crewmen can be examined from present unmanned satellites, and the image provided lacks the precise definition supplied by the crewmen."

In a summary report, the scientists put it this way: "Man sees correlatively, observes and analyzes relative to his own knowledge, selectively observes, integrates what he sees into a 'permanent' record and updates that record, compares what he observes with the record, and makes conclusions based on this comparison." Humans and sensor packages, the scientists concluded, "are each uniquely adapted for specific functions and . . . these functions are, in many cases, complementary."

The task of measuring color is not an easy one. Skylab had a rudimentary reference book. On Salyut-6, the library included a color atlas with 192 hues. This was totally inadequate (the eye can distinguish almost a hundred thousand different shades and hues), and a bigger volume, containing more than a thousand plates, was sent up later. But even this proved insufficient for precise and distinct identifications of all the colors the cosmonauts were seeing.

The Soviet solution to this problem was a small device called the Tsvet-1 ("Color-1"). The size and shape of a camera, the apparatus has three control levers that can be manipulated to alter a color display on a small front screen. By 1983, cosmonauts were using the box, called a "colorometer," to quickly find any color they were seeing; the lever settings were then read down to Earth where duplicate devices, plus calibration equations, determined the precise brightness and wavelength.

To calibrate color observations, Soviet scientists in 1983 spread a vegetable dye in the Black Sea, near Odessa, to form a spot about half a mile across. The crew easily saw the spot and measured its color.

Visual observations were not limited to colorations of a flat surface. Spacefarers clearly saw differences in elevation, even on the ocean! Some cosmonauts observed "steps" on the ocean surface: "We see ledges. From here it's difficult to determine, but sometimes they are as much as ten meters," one team reported. Kovalyonok claimed to be the first Russian to see them, in 1978; he also mentioned a trough near the Caroline Islands, and a swell up to one hundred kilometers long in the Timor Sea. These ledges had in fact been first seen by Vance Brand in 1975 during the two-week Apollo-18–Soviet linkup mission.

Kovalyonok claimed that he could see the tiniest details, even submerged islands that do not show up on film. Another cosmonaut reported that his rookie shipmate's pleasure was "to gaze on the mountain crests beneath the sea, if he is lucky (scientists are still cudgeling their brains to find out why these are visible)." Skylab astronauts, too, had seen those subsurface features, particularly the reefs and shoals south of Ulithi Atoll, the French Frigate Shoals, and near Fiji and Tonga. One attempt to detect Orne Seamount, thirty meters below the surface, proved fruitless.

Sometimes, familiar objects were seen from such a new perspective that their appearance was entirely different from what it was when observed from Earth. Spacefarers frequently caught glimpses of comets, meteors, passing satellites, and the northern lights. It startled them to see the "falling stars" below them.

On Skylab-4, the three astronauts used visual observations of the Kohoutek Comet to provide supplemental data on structure, brightness and color. No camera on Earth or in space could record such fine detail. The astronauts experimented with techniques of dark adaptation, used spare time to perfect the techniques, and compared notes with shipmates. As Pogue recalled, "Many of the subtle features of structure and color were determined by consensus after crew discussions and follow-up observations." The men had made two spacewalks, on December 25 and 29, to aim cameras and to make eyeball observations. Ed Gibson, the physicist on the crew, later provided the most detailed description of the comet's size, orientation, tail color, and prominent spike that stretched out toward the Sun ahead of the comet in its path. NASA's official Skylab history declared: "Skylab's observations of Comet Kohoutek were a small part of the total study, but they were among the important ones. . . . The crew's visual observations and color sketches were far better than any such made from the ground."

In recognition of the scientific quality of these observations, both American and Soviet manned spacecraft will take part in the Halley Watch early in 1986. Certainly a crew (perhaps including an astronomer) will be aboard the Salyut-7 space station; a special shuttle mission dubbed Astro-1 will be in orbit with four astronomers aboard.

On the night side of Earth, spacefarers have been able to observe stars down to the fifth magnitude in brightness, not quite as good as from atop a mountain on a moonless night on Earth.

In 1978, Georgiy Grechko made extensive sketches of astronomical phenomena, including the rising of Venus, the aurora, the setting of the Sun past a cloud-free horizon, and noctilucent clouds in the ionosphere. He became such an enthusiastic observer that he got into trouble with Mission Control: convinced (correctly, as it turned out) that he had seen reddish glows in certain bands of the ionosphere, which correlated with solar activity, he conducted an after-hours observation program (he even recruited his crew commander to assist him), and was twice reprimanded by the doctors for his "adventures" that were interfering with his sleep. On later Soviet missions, similar observations were officially scheduled.

Astronauts on Skylab in 1973 had also gotten excellent views of the aurora borealis. Jack Lousma spotted it first, and called the others to the window. Owen Garriott, an ionospheric physicist before becoming a scientist-astronaut, described it: "It was greeny in color, I couldn't see any red. The arc extended out away from the spacecraft toward the horizon. . . . Above it, very faintly, you could see streamers, very thin striations extending from much higher altitudes—very thin rays, very dim but thin rays more or less vertically aligned. . . ." No photograph ever captured this view.

Ten years later, cosmonaut Aleksandr Aleksandrov tried to put his own visual impressions into words very similar to Garriott's: "Well, to give you an idea, you could say that on the mid-horizon there are columns with a green glow and from each one of those columns there is a winter road leading away, with snowdrifts, winding towards you." Don Lind's most memorable impression of the Aurora Australis was as it rippled from horizon to horizon, below his vantage point aboard Spacelab-3. Visible below the pastel ripples were moonlit clouds. Across the gap of technology and politics American and Russian spacefarers shared these common experiences.

It's not only the long-duration spacefarers who have kept a sharp eye on Earth. In 1983 shuttle astronaut Norm Thagard, an M.D. and former Marine Corps carrier pilot, told a news conference about his favorite way of viewing Earth. "I turned myself upside down over in the right-hand seat—and we were usually with the Orbiter upside down looking down at the Earth. I remember the pictures of the old blimps and their gondolas—and the picture is, you could just sit there and look at the world go by. That's exactly what I did. It was like being in the gondola of a dirigible, only this dirigible was a hundred and sixty miles up in the sky. It was great."

For two decades, most eyeball observations were incidental to other activities. Early spacecraft windows were small, and crewmembers were rushed and preoccupied except for an occasional "Oh, what a beautiful view." But gradually, as missions lengthened and as scientists came to realize the value of eyeball reports, more time was spent in serious observation. Ultimately, late in 1984, the first pure observer was sent into space. His job was entirely dedicated to looking out the window.

Paul Scully-Power, an Australian-born oceanographer, had been analyzing space pictures and debriefing spacefarers for many years. Even at second hand he and his associates could perceive many tantalizing optical features. So when a spare seat was opened on a routine space-shuttle mission, he was selected to go—at three months' notice. For scientific equipment he packed a battery of cameras, tape recorders, maps, and notebooks. His goal: to make "one major discovery" looking out the window.

In orbit, he stayed at the windows every waking, sunlit moment, using the periodic half-hour "nights" to take care of practical matters. Like Thagard he used "the gondola position," upside down in the front pilots' windows: "It's by far the best way of looking at the Earth."

"The most spectacular thing I saw was the Mediterranean with its dynamics," he summarized. Spiral ocean eddies, first really noticed by Skylab astronauts more than a decade before, covered the entire body of water. Moreover, Scully-Power discovered, "they were interconnected over the whole length and breadth of the Mediterranean. . . ." It was, he concluded, "just a tremendous observation."

In his week in orbit, the oceanographer observed another correlation that had escaped the notice of earlier spacefarers: ocean currents often were manifested in cloud structure over them. "At one point, I became aware of linear breaks in the clouds, which were running east to west. . . . [It] was about sixteen miles wide and went east-west as far as the eye could see, and there was another break farther south." The expert knew what must have caused the visible features: "It was in the Pacific, and it turned out to be the equatorial counter-current that caused the first break, and the south equatorial current was the second break." A temperature differential in ocean water, he realized, "will have a line of clouds associated with it."

A drawback of such short-term observations is that the sharpness of visual observations does not seem to begin immediately. On Skylab, the men seemed over time to "learn tricks" for better viewing. Salyut spacefarers found that it takes several weeks to become fully acclimatized. Late in 1983, cosmonaut Vladimir Lyakhov, veteran of an earlier six-month mission and a recently concluded five-month mission, explained it this way at his postflight press conference: "I should point out that there has to be a period of adaptation, before you can start making serious visual observations of one phenomenon or another, because when you first find yourself in orbit, your eyes are simply not yet accustomed and you are not psychologically ready to perceive what you see. This period, for me, lasted something like one month."

Elsewhere, he had remarked: "Although the ocean's surface seemed at first to be monotonously homogeneous, after half a month we began to differentiate the characteristic shades of one sea or another and different parts of the world ocean. We were astonished to discover that during a flight, it's as if a cosmonaut learns how to see all over again. At first the finest nuances of color elude you, but gradually you feel that your vision is sharpening and your eyes are becoming better, and all of a sudden the planet spreads itself before you with all its unique beauty."

Civilian engineer-cosmonaut Valeriy Ryumin reported similar experiences: "It's not easy to find Earth reference points," he told an international astronautics conference in 1981. "For example, during my 175-day flight [in 1979] it took me one to two months before I could detect color anomalies in the ocean and send the information back to specialists. But during the 185-day flight [in 1980] I could send such information back by the end of the first week."

This phenomenon of rapid recovery of "space sight" appears to be common. In 1982, Anatoliy Berezovoy spent seven months aboard Salyut-7. Early in the flight, the station was visited by another short-term crew, which included some space veterans. Berezovoy recalled that one of them "remembered nuances in the Earth's surface quite well, and taught us how to observe various phenomena in the sky. He showed us zodiacal light. . . . We observed the Aurora Borealis for the first time with them."

This perceptual progression thus appears to be a learned capability, related to accumulated experience and on-the-scene advice. "During the five months of flight, we spent a total of nearly a thousand hours close to the portholes, carrying out visual observations," noted Lyakhov in 1983. That comes out to about three hours per man per day.

But Paul Scully-Power believes the learning curve can be traveled on Earth as easily, if not more easily, than in orbit. He spent years poring over space photographs of Earth's surface, and when he went into orbit himself in 1984 he reported seeing sharply and clearly—as sharply as the Soviets or the Skylab astronauts ever described their seeing—from the very first day.

From the earliest days of manned spaceflight, some spacefarers have returned with claims of extraordinary visual acuity, of seeing individual ships, trains, pipelines, buildings, and other relatively tiny objects. The claims were met with incredulity by medical specialists, who computed that they were far beyond even the theoretical sharpest angular resolution of the human eye. Subsequent experience has been that while eyes in orbit can work wonders, they cannot work miracles.

Astronauts have reliably reported seeing ships' wakes, but claims of seeing ships themselves have been dismissed as subconscious suggested imagination. Soviets had the same debate. Viktor Glazkov claimed to have seen a bus traveling along a road in Brazil, but doctors dismissed it as "an illusion produced by mental associations." Vitaliy Sevastyanov thought he could make out a school of fish in a bay.

Under a special observation program, cosmonauts have actually performed visual near-miracles. In 1978, two spacefarers detected a thirty-meter crater in a glacier, and correctly estimated that it was about two hundred meters from the end of the ice flow—but that was under absolutely optimal conditions of contrast and illumination.

Such disputes—and the claims that have already entered spacefarer folklore—have little practical significance, since people in orbit can and do carry artificial viewing aids, binoculars and other magnifying equipment. More important than the power of the lenses used, however, is likely to be the quality of the window.

Poets have called the eyes "the windows of the soul." But spacefarer poets may well have a different metaphor: their station windows are "the eyes of the soul."

Mercury astronauts had to fight to get a window in their spacecraft; on Vostok, the small porthole was between the cosmonaut's feet. Gemini had two tiny half-moon-shaped windows placed in front of both crewmembers' faces, but the angular view was no larger than two hands held at arm's length. It wasn't until Apollo and Soyuz, in the late 1960s, that windows more than six or eight inches in diameter appeared. And then crewmembers had time to float over to them to look out.

The Skylab window, whose very existence was the subject of ferocious and drawn-out debate, was a foot and a half in diameter. In flight, it was described as constantly smudged with nose- and fingerprints (since it was cold, the crewmembers would rest their foreheads against a piece of cloth on the glass). On Salyut, there were several small windows in the airlock module, at which instruments could be mounted, but the biggest for sightseeing was still only about six inches across.

By comparison, the shuttle is a virtual flying greenhouse. From the pilots' seats in the nose, four contoured windows give a 180-degree wraparound field of view (this is where the tourists go "gondola mode"). The roof of the flight deck also has two large windows, and there are more windows facing aft into the cargo bay. Downstairs there are small windows in the airlock hatches (looking aft), and in the entry hatch on the side of the ship.

In general these windows (on both Soviet and American spacecraft), and spacesuit helmet visors as well, mimic one condition of the atmosphere that scientists tried but failed to overcome: although they are opaque to damaging solar radiation, such as ultraviolet light, and a human being can look out the window in safety, in exchange, Earth's wavelength blindness has been artificially reproduced.

Some observations require working at wavelengths outside the safe human range, so one window on Salyut and one (the downstairs hatch) on most of the space shuttles have been built of UV-transparent material, such as quartz or fused silica. An additional protective cover is usually installed, which can be removed when such photographs are needed.

Accustomed to the benign conditions of Earth, where sunlight is dangerous only when concentrated (such as arctic snow blindness) or extended (while sunbathing), some naïve spacefarers have risked injury at such windows. In 1982, one cosmonaut received a mild sunburn in the few moments it took to mount a solar camera. In 1984, doctor-astronaut and Skylab veteran Joe Kerwin had to issue a stern memorandum on the hazard of the new shuttle hatch windows.

According to Kerwin's caution, the danger was very high: "Five seconds' exposure to solar radiation through the window will produce a perceptible effect; ten seconds' exposure will have severe consequences . . . [and] may cause irreversible damage to the cornea." Direct exposure was not even necessary; standard assumptions about reflectivity showed that fifteen minutes' exposure anywhere in the middeck "could yield severe effects." And even if the window is turned away from the Sun, solar ultraviolet could enter after being reflected off Earth or off the spacecraft structure outside.

The best view of the outside is from the outside. Astronauts and cosmonauts who have walked in space (with UV-protective visors, to be sure) have testified that the view is markedly superior to even the best from inside.

"Immediately after the airlock was opened," said Anatoliy Berezovoy, "the impression was that of being on a street on a bright sunny day with the ground covered in pure white snow." To Georgiy Grechko in 1977, the sky seemed more colorful and vast, and he could feel the Sun's heat even through his suit. "It really is very impressive," Aleksandr Aleksandrov recalled about his 1983 experience. "And the impression is a lasting one. Perhaps for the first time in the whole flight, we saw the dawn through a very good glass—I am thinking in terms of field of view and light transmission—in the helmet of the spacesuit. We saw through virtually 360 degrees of the Earth's surface. This indeed was a unique sight, and we would not have been able to see it out of any porthole."

Jack Lousma had an even more graphic comparison to describe his spacewalks from Skylab in 1973: "When you're inside looking out the window, the Earth's impressive, but it's like being inside a train; you can't get your head around the flat pane of glass. But if you stand outdoors, it's like being on the front end of a locomotive as it's going down the track." In angular size of the scene, it's like the difference between watching television and sitting in the front row of a wide-screen theatre.

A perfect orbital observatory would have many of the features described by Soviet and American spacefarers, both in terms of field of view ("as good as outside") and transparency. First would be an unrestricted view of up to an entire hemisphere—perhaps a transparent bubble protruding from the station's earthside bulkhead.

A control facility in the observing bubble would provide for manual or automated pointing of instruments. Some sort of indicator would perform surface feature tracking, and within view would be a moving map display showing the area in sight, in perspective. Previous views of the same area, hours or weeks or years earlier, would be viewable as well, with perspective corrected to that of the current pass (this will require considerable computer power). The observer would be able to direct a pointer wand or scope sight at any feature, and receive an immediate readout of location and geological/ cultural identification. The observations would be logged and time-tagged along with spoken commentary.

Hand-held optics would be essential. Even on Skylab, the crewmembers used their 10x50 binoculars at least half the time they were at the windows. On the shuttle, crewmembers reported that with their 10x binoculars, or with the 250-mm lens of their 70-mm camera, they could see ships of destroyer size, but could not identify their exact type. The 10x glasses were deemed the most powerful practical for hand-held use; gyro-stabilized viewers were suggested for several times the power. There needs to be an easy way to switch between magnified view and normal view, to allow orientation and target acquisition.

Zoom optics would be available, with variable magnification and field of view. Control parameters would be displayed "heads-up," projected against the viewport itself, so observers would not have to repeatedly switch gaze and focus back and forth. A real-time image repeater/monitor would allow the observer to watch exactly what the instrument was picking up; the signal could be false-color enhanced, and superimposed image or map/graphic combinations could be created and instantly displayed and studied.

What would an observer look for? Judging from the things spacefarers have noticed in the past, the most productive role of an earthwatcher would be to keep an eye out for the unexpected. This is the kind of instruction that cannot be followed by mechanized instrumentation. Only humans can understand and follow such orders.

Abnormal colorations, streaks, light and dark patterns, and so forth are the kinds of visual stimuli that led to some of the major visual discoveries already described. There is no reason to change the game plan that works so well.

When Paul Scully-Power was in orbit in October 1984, he noticed something that should have been a hint about Earth's climate—but it wasn't followed up. "I am not a meteorologist," he explained, "but I am pretty certain there was anomalous meteorology going on, for two reasons. The first thing I realized as soon as we got on orbit was that there was far more cloud cover around than I expected. My guess is that at that time of year you could look at the total Earth and you would expect about thirty percent cloud cover. I would say we were looking at more like sixty or seventy percent cloud cover while we were up there, throughout the whole Earth, and that surprised me." Second, there should have been overcast differences between the northern and southern hemispheres, but there weren't.

So the productive strategy, based on these experiences, is clear: assign people familiar with the workings of Earth's oceans and atmosphere to sit by the windows and watch the planet. Every once in a while, ask them if they see anything strange. Earth will guarantee they always will; wisdom will guarantee we'll be better off knowing it.

chapter five

NO EARTH IS
— an —
ISLAND:
Keeping the World
Habitable

To paraphrase Rudyard Kipling, they little know of Earth who only Earth know. . . . It isn't that in going out we ignore the Earth, it's in going out that we can get the proper attitude required to save the Earth.—*Isaac Asimov, 1977*

We have looked at the Sun for years, and the Moon, but we have never really looked at the Earth before, and that shouldn't be. . . .
—*Ed Gibson, Skylab astronaut, 1973*

Physicist Jack Paris, an expert in remote sensing from space, strolled through the warm-steamy, hot-desiccated, cool-shady atmospheres of the Los Angeles Arboretum's vast collection a couple years ago, getting to know the individual plants in his unique way. Already an expert in crop vegetation, Paris was alert to each specimen's leaf size, each leaf's orientation to its stem, and each plant's chlorophyll signature—the unique way it absorbs light and leaves its mark on observing instruments.

Such a sophisticated manner of observation was impossible twenty years ago. There were no disciplines of knowledge called "Earth Resources" or "Remote Sensing." There were, of course, departments of geology, oceanography, botany, agronomy, and so on, but these were separate and discrete areas of learning. Each looked at Earth with a narrow, specialized perspective.

The unthinkable has now become the obvious. The conceptual revolution in considering Earth—its past, present, and future—came about through the technological revolution in actually viewing Earth as a planet. The view from space via the eyes and the subtle senses of instruments gave these discrete areas of knowledge a larger perspective and new tools of observation. Most important, specialists began to appreciate how Earth works as a dynamic, adaptable system, greater than the sum of its parts. They glimpsed the relatedness of their areas, and tried to think of means of speaking to one another and using the new space technology in their work.

The earliest applications of this new learning were directed toward crop predictions. Specialized Earth-observation satellites, especially the early Landsats, harvested new, detailed data with their large number of optical and infrared sensors. These could detect disease and insect infestations in key crops. Additional applications for these painstakingly detailed, digitally processed pictures followed. Surveys of third-world nations rendered pictures of massive clear-cut areas. Radar pictures on later Earth resources survey spacecraft even penetrated beneath the surface of the Sahara to reveal a prehistoric riverbed along which archaeologists hope to find human and prehuman relics.

The early Earth resource satellites also detected phenomena that had hitherto been only poorly appreciated, if at all. They whetted the specialists' intellectual appetites for a more comprehensive understanding of Earth—its climate, its manner of keeping its climatic balance, man's true impact on the planet, and ways in which environmental disasters can be assessed, predicted, and averted.

In 1982, a group of fifty scientists from various disciplines

gathered at the Woods Hole Oceanographic Institute to artic-
ulate the need for an interdisciplinary research program de-
voted to long-term study of global change. While humanity's
food, air, and water needs were met in the past by exploiting
resource-rich frontiers and developing advanced agricultural
technologies, the group stressed that "Now humanity is being
confronted by the finite dimensions of its world." With seem-
ing irony, they concluded that only by going beyond the
boundaries—physical and conceptual—of Earth, out into space,
could the Earth's contents be adequately measured, mapped,
and managed.

In the Earth's 3.5-billion-year history, life forms of all kinds
have been threatened, mainly because of destruction of their
habitat, their food, water, and air supplies, by climatic change.
Most have lost the struggle. No one had even an inkling of
the rules of the game—until now. "The human species faces
a similar challenge," the scientists wrote. "In contrast to its
less successful predecessors, however, humanity has the ability to manage its resources, to plan intelligently for its future,
and to preserve the necessary elements of its habitat." Only
the space perspective offers this opportunity.

Climatic change can be brought on by cataclysmic events,
such as volcanic eruptions or advancing ice sheets, which
alter temperature and circulation patterns. On a smaller scale,
the human race can influence the chemistry of the atmosphere
by altering ozone levels with trace contaminants, or by burning fossil fuels and clearing vast forests, both of which raise
carbon dioxide levels. Soil erosion, desertification, deforestation, overgrazing, squandering of freshwater resources, acid
rain, air pollution, and diversion of rivers also influence global
habitability, the conditions on which all existing life forms,
including humankind, depend. With a glance, spacefarers
noted precisely these phenomena where their scope had never
been appreciated before; with highly instrumented automatic
space platforms, the degree of these kinds of damage can be
plotted, and trends over time measured and projected.

Doomsday predictions of climatic disaster as a result of human activity have been made often and loudly, but they have generally been based on inadequate data and knowledge of how the Earth works. The Earth is adaptable; it is used to variations, and it has a surprisingly flexible array of compensatory mechanisms.

For instance, rising concentrations of carbon dioxide in the atmosphere act as invisible fertilizer. Scientists at the University of Arizona's Laboratory of Tree-Ring Research found that certain trees growing in high altitudes had increased widths in their annual rings in recent decades. Whether this is a result of higher carbon dioxide levels alone, or of warmer weather and greater moisture (which might also be a result of higher CO_2 levels) is not known. The plants flourish, release more oxygen into the atmosphere, and respond to this change by a natural compensatory reaction. Just how far these compensatory reactions can be pushed before we head into a greenhouse effect or a new ice age is unknown. Earth does have certain pressure points, areas of its biosphere, where even small changes can have enormous consequences.

In 1983 and 1984, a number of scientists spent a great deal of time thinking about just how much we do not know about the Earth's systems. In order to design new tools of knowledge, they first had to plumb the specific depths of our ignorance.

All life on Earth depends on water, on the dynamic, restless metamorphoses among its snowy, fluid, and vaporous forms. Mechanisms involved in water's ever-changing state are tied to air circulation and the redistribution of energy and heat from one part of the planet to another. The evaporation of water from soil and its transpiration from leaves into the air is essential to plant growth—it keeps plants from overheating, and provides a source of future precipitation, rain that will nourish other plants.

Only 3 percent of Earth's water is fresh. It is vital to terrestrial plant and animal life, but most of it is locked into the polar ice caps and glaciers and in ground water. Snow cover and snow melt are major components of the world's total annual water supply; a frozen-water inventory, which would be impossibly expensive if performed on land, can be accomplished easily by a few orbital overflights.

Only crude estimates of the amount of water stored in the poles, soil, water vapor, and fresh-water lakes and rivers exist now. What effects soil evaporation has on food production, air circulation, and global water supply is unknown. Rainfall is still inadequately measured by rain-gauge networks and meteorological radar. The nature and dynamics of snow cover, especially its extent, depth, water equivalent, wetness, and albedo (reflectivity, or degree of whiteness) are not understood; in some cases they are not even quantified. Precipitation and evaporation over the oceans are unknown within at least a factor of two.

Instabilities in the polar ice sheets could have a devastating effect on sea level and climate, but we lack an understanding of the interaction between the ice sheets and the oceans, especially the processes involved in melting and refreezing of ice walls and beneath ice shelves.

Coastal zones around all the continents play an immense role in cycling fundamental life chemicals such as carbon, nitrogen, potassium, and sulfur. The bulk of the world's harvest of fish is found in these zones. Past fluctuations in climate may have been partially triggered by how the ocean produces and cycles marine life and important atmospheric gases. Still, current knowledge of how the biosphere, atmosphere, and ocean interact remains sketchy.

Huge forest areas such as the Amazon basin provide a large amount of oxygen to the planet, and are important recyclers of water, carbon, and nitrogen. Major deforestation efforts are underway in the Amazon and other vital forests, which may have a profound effect on watersheds, soil erosion, planetary energy balance, and regional climate. Scientists would like to know how much biomass is lost over various periods of time, how much of it winds up in rivers and is then emptied into the oceans, and how much and what kind of regrowth occurs after the initial forests are cut down.

Variations in plant growth in the ocean, especially the abundance of phytoplankton, must be charted, since phytoplankton lies at the base of the aquatic food chain and may play a large part in other geochemical processes.

These are only the biggest gaps in our knowledge of the planet. Others having to do with how the sea interacts with the air both physically and chemically, how the various layers of the atmosphere play their separate roles in the maintenance of life, how large geological events involving plate tectonics affect life forms and climatic processes have their unknowns too.

How do we begin to study something as complex as the Earth, with all its systems interacting simultaneously and fluctuating over years, decades, centuries, and millennia in accordance with changes in its own system and extraterrestrial events such as solar activity?

The Earth observation satellites and weather satellites were a good start. Weather satellites in polar orbit and at geosynchronous orbit produced accurate pictures of previously unperceived large weather systems. Time-lapse photography gave graphic depictions of large weather system behavior. The Landsats enabled scientists to predict crop yields with 95 percent accuracy. They were also crucial in mapping and inventorying many poorly surveyed areas. Topographic hints of underground mineral deposits were spotted. With these satellites, soil, sediment, and rock characteristics of many areas were now seen, even through dense vegetation. Additionally, the Seasat satellite and its imaging radar system enabled the mapping of polar areas and a better understanding of the role of the poles in climate; Seasat also contributed to oceanic geophysics by plotting gravity variations manifested in sea surface characteristics.

Although Landsat and other Earth resources satellite photos were useful, there were many measurements that they could not carry out. "You can't tell the difference between wheat, barley and rye on Landsat," Jack Paris pointed out. "They are very similar to each other since they're all grass crops. Landsat can tell forest from nonforest, but it can't distinguish between forest and areas where regrowth has occurred. It uses a vibrating mirror that looks back and forth across a one hundred meter area, and doesn't look at one area for a long time." Each Landsat picture is composed of thousands of tiny squares, processed in the slow and laborious digital manner. It takes six expensive hours of data processing to produce one picture. Some scientists are allotted only one picture every three or four months because of the processing problem. As a result, only about 5 percent of Landsat's pictures have ever been put to use.

The Landsat program, however, was under budgetary attack throughout the late 1970s and early 1980s, and the U.S. government considered selling it to a private company that would administer and distribute its data. The program was costing $100 million a year to operate, and was selling only about $15 million worth of products. However, the sell-off effort met with strong opposition, and was later squelched. The launch of Landsat 5 in March 1984 was the last of the program.

There are other useful "remote sensors" in orbit. N-ROSS, the naval research oceanographic satellite system under development, will carry five major instruments to study the ocean. It is due for launch in 1988. The following year the TOPEX (Topography Experiment), which will look at ocean current signals and ocean topography, will be launched. That same year the Upper Atmospheric Research Satellite (UARS) will be sent up to study solar effects on the atmosphere, and take global measurements of nitrogen and chlorine ozone.

Except for some smaller remote-sensing projects, such as the short-term SIR (Shuttle Imaging Radar) series (A and B already launched and C due in 1988), and some limited experiments aboard a few Spacelab flights, nothing is in store for remote-sensing scientists and their great dream of understanding global habitability. This is not to disparage the important projects underway, but a vast, unified "big picture" is not in the works.

SIR-B had been slated to search for an ancient lost city in the Saudi Arabian desert, but antenna problems scrubbed those runs. However, a remote-sensing expert sitting at his desk in Oklahoma City found an entire gaggle of lost Mayan cities in southeastern Mexico just by looking at existing Landsat pictures in a new way.

Rod Frates, one of the founders of a consulting firm specializing in interpreting Landsat data for the oil and gas industries, had been reading an article in *National Geographic* about the difficulties encountered by a team of explorers hacking their way, on foot, through the Central American jungles. "These guys sure are going about it the hard way," he recalls thinking. "I knew something about remote sensing and I just thought that satellite imagery was a good way to become an instant archeologist."

With a small grant from a research foundation, Frates and a team of associates did exactly that. The Mayans had excavated reservoirs as large as ten acres to hold fresh water, and even after almost two thousand years the vegetation grew less densely in the dried-up remains—they showed up as distinct dots on properly encoded Landsat images. Frates found a number of cities (including one, Oxpumel, which had been stumbled on by lost explorers who then couldn't find it again), and a vast array of walled fields that were unknown to archaeologists.

In November 1983, Frates and a helicopter team flew over twenty of more than one hundred "possible sites." At each site, data from the satellite-based "Global Positioning System" allowed them to determine their location to within twenty feet. At one site, Frates was lowered by rope to the top of a pyramid, where the surrounding underbrush was too thick for the helicopter to land. He took a chain saw and began cutting a clearing, but the saw broke. Before he got hauled back up he surveyed the pyramid—probably unwalked on by human beings for more than a thousand years—and a large plaza that led to another pyramid half a mile away. All around the pyramids were smaller buildings. The data were turned over to Mexican archaeological authorities for further exploration.

For more conventional applications, some governments have planned Earth resources missions of their own, although none is as comprehensive as Landsat. The French SPOT (Système Probatoire, d'Observation de la Terre) satellites will survey land use and do resource inventory work. A series of marine- and land-observation satellites are being launched by the Japanese. The marine satellites will study color and temperature of the sea's surface with Seasat-type spaceborne radar technology, while the objective of the land satellites is geological survey, land use, forestry, and disaster prevention. The Canadian Radarsat series, involving three or four satellites, will try to conduct high-resolution studies of arctic areas, since it is in Canada's economic interest to navigate through its frozen northern areas to oil-rich fields. Ocean studies, agriculture, forestry, and water resource management will also be included in the instrumentation. Finally, the European Space Agency (ESA) will launch a remote-sensing satellite system in 1988 or 1989 to study land use and global weather.

International cooperation has always been considerable in remote sensing, but none of the projected efforts comes close to the type and magnitude of data necessary to study the Earth in its totality.

Remote-sensing scientists approached the problem sensibly. The first move was a call for putting Landsat data to fuller use, and creating a centralized Earth resources data base. The latter was essential, since such data are currently scattered and almost inaccessible. A data base, the scientists felt, would serve as a foundation on which to build and store future knowledge. A centralized library, a storehouse for all we know about the Earth—from laboratory, ground and field experiments, aircraft overviews, and past and future satellites of various types and in various orbits—might prove to be the most important body of knowledge ever collected for the benefit of the human race.

Apart from a data base, sometimes called an "archive," these scientists hungered for better and more refined remote-sensing equipment. When President Reagan announced in his State of the Union message in January 1984 plans to build a space station, the remote-sensing scientists took heart, as did most of the space community. They had been dreaming of an ultimate and permanent remote-sensing system that would not be under constant budgetary threat (like the Landsats), and could carry out uninterrupted research for a long time. Studying the Earth, they all agreed, required a system that built on current data and carried out investigations for twenty to fifty years so that natural cycles could be witnessed. Thus "System Z" was born.

It is unclear just how System Z got its name. One scientist close to the project thought it was simply one high-ranking individual's happy expression of what "the ultimate" remote-sensing project should be named. And for a long time everybody called it that, until NASA found a less sporty, more technical-sounding name for it: Earth Observing System (EOS).

Once a space station became a national commitment, many other projects became possible. But with a space station, the shuttle, and all the peripheral satellite servicing equipment that will be developed by the early 1990s, it will be possible to design a permanent remote-sensing satellite that can grow and evolve.

System Z can be described as a polar orbiting platform that could be serviced and perhaps even controlled from the ground, the space station, or the shuttle. Broken equipment could be detached and retrieved by teleoperated robots sent up from a polar-orbiting shuttle mission (with a small space tug on line, it would even be possible to have it tended by visiting humans). New and updated equipment could be added to the instrumentation, provided System Z is built with this desirable flexibility.

It would be wasteful for System Z to duplicate the capabilities of the existing satellites. Instead, current thinking is to install instrumentation that will fill in the gaps in our knowledge of essential areas of the Earth's system. For example, among the plausible instruments for System Z are some with very high resolution, that is, instruments capable of focusing sharply on very small biological systems, in order to give detailed information on the type, extent, and state of biological activity on various sites on Earth. Variations in grassland and mountain areas would be easy to see with such an instrument.

The instruments will be designed to work synergistically, that is, to work so that the complete body of knowledge is greater than the sum of its parts. For instance, spectral instruments can measure plankton growth in the oceans, ocean circulation, and some aspects of water-air interactions. Microwave radiometers measure water and energy balances of the Earth's surface; LIDAR provides information on how water vapor is distributed in the atmosphere. Flown together, they measure and quantify important aspects of global hydrologic cycles. Over land, these instruments can help determine the range of different ecosystems, types of existing vegetation, and the state of the vegetation.

Some space sensors will work with instruments placed on the surface of the land or ocean. Ocean buoys and below-ground soil probes will be part of the Automated Data Collection and Location System (ADCLS). Soil probes can measure moisture with great accuracy, and buoys can measure things like imported nutrient cycling between the deep ocean and surface waters. This network of buoys and probes would relay data through the satellite. A comprehensive overview of even such minute processes could be mapped, as it were. That is, very local, specific data can be combined into a large and detailed picture of these dynamics.

Having learned the lessons of Landsat, the scientists first

applied themselves to the problems of data management. Much of Landsat's capability was wasted due to cloud cover and to its inability to throw away useless data. System Z may use "smart sensors." Data could be preprocessed to remove transmission errors, and could downlink essential information whenever it's commanded to do so. The computer systems and data processing models have to be worked out to achieve this; it is a complex task for an automated system to process information intelligently.

Another idea is early use of artificial intelligence to help in the preprocessing procedure. The Intelligent Earth-Sensing Information System (IESIS) would endow the data-gathering system with a detailed "world model," some language capabilities, and limited decision-making abilities. IESIS would carry a description of the Earth's topography and environment, and could be asked to observe closely specific features. Unnecessary and duplicated information could be dumped automatically. It could also identify anomalies, sensor readings differing from its preprogrammed world model, and could then downlink all sensor data for human analysis and identification.

Because System Z will bury scientists under an avalanche of new data, the scientific community will have to develop the intellectual tools to make sense of it all. Undoubtedly, mathematical models of the Earth's complex and interrelated processes will have to be worked out, and changed and refined as the system proceeds. Even if IESIS were someday to be used on System Z (there is no discussion of its use early on in the operation), the knowledge base would have to be worked out very carefully ahead of time by teams of experts, and then fine tuned over the years.

A new space platform is not enough. Not all remote-sensing needs can be met in System Z's near-polar/near-Earth orbit. System Z is only one powerful and evolving tool that will be coordinated with other specialized satellites, and with essential earthside studies. These studies will always be necessary to obtain "ground truth," to double-check the accuracy of space data and to calibrate the ever-changing technology aboard the satellites.

For instance, although weather satellites give excellent pictures of, say, an evolving hurricane, the pictures are only part of total hurricane data. Specially equipped airplanes are still flown through a hurricane to get specific data on wind speed, barometric pressure, circulation, and steering currents. Weather stations on shore monitor the storm with radar, finding its pockets of violent activity. As a hurricane approaches a shoreline, points along the shore note local barometric pressure since hurricanes sometimes take the path of least resistance, that is, pick a place with low barometric pressure to make landfall. While a satellite alone could not possibly provide all the necessary data to help coastline residents decide whether or not to evacuate their homes, it is an important piece of a large and complicated puzzle requiring ground, air, and space efforts for even approximate predictions.

If global habitability is the aim of this large and ambitious project, it must be international in its scope. Ground measurements of tropical forests, coastal shelves, and major estuaries cannot be taken without the permission and participation of many countries. Some international programs already exist for this. The Global Atmospheric Research Program (GARP) is a good example of a program in which large and small nations sought to set up a large observational system. The same is true of the World Climate Program, whose job is to assess the impact of climatic changes on agriculture and water supplies.

A large, all-encompassing program could make a wall-to-wall inventory of Earth's resources, although it would be costly. Suggestions based on detailed data analysis could be made to various locales to help them understand water allocation. Localities within developed countries might prohibit building on prime, arable land, and instead steer development to less tillable properties nearby where, perhaps, the soil has too much clay to allow for good crop growth. Program scientists could, if they were sure, ask a country not to undertake a particularly environmentally destructive project, or to modify it so it is less destructive to the world's peoples. This would be hard to police, although in fifty years or so environmental pressures on our species might be so severe as to make such suggestions diplomatically, economically, and politically compelling.

"The task is urgent," wrote Richard Goody in the executive summary of the 1982 Global Habitability Workshop at Woods Hole. "Because the human race has reached the point of being able to change the Earth significantly on a time scale of mere decades, its ability to mount countermeasures, should it wish to, is restricted in many cases to a time scale that is similar or longer. . . . [T]here is an obvious need for international cooperation. We can see no better use for the mastery of near space than the acquisition of the body of knowledge essential to the future well-being and prosperity of humanity."

Ultimately, the tools we develop to study our planet may give us insight into others, as we push out into the Solar System. For instance, Mars has complicated atmospheric dynamics, although not as complicated and dynamic as Earth's. On Mars, so far as we know, water exists only as ice and fog, not as open water even though there are many features on the planet that suggest the existence of open water at some time in its history. There are seasonal variations, cold fronts, storms and such, although they differ greatly from those we are familiar with on Earth.

However, some instruments used to study Earth can be used to study Mars, and another library, founded on Viking data, may be started to understand Martian dynamics. Instruments aboard the probe to Mars to be launched in 1990 will be similar to those on the later Landsats. These and others developed for System Z may help us probe the very different regions of Mars, its atmosphere and, in the unlikely event we find some form of life on Mars, its life forms. Armed with that knowledge, we may be able to understand not only what makes Earth habitable, but what made Mars uninhabitable.

The insights we gain into other planets will come home to roost. How can we expect to understand any planet, even if we're standing on it, if the size of our sample is one?

The conceptual revolution goes at least as far back as 1968, when photographs of a full disk Earth, taken by moonbound Apollo astronauts, electrified our consciousness and gave major (some think crucial) impetus to the environmental movement. The revolution continued when planetary catastrophism, a taboo topic discredited by Velikovskian crackpots, sprang up invigorated by discoveries and observations on a dozen worlds, and almost overnight the "asteroid catastrophe" theory for the extinction of the dinosaurs 65 million years ago received near-universal scientific respectability.

The revolution has come home with "nuclear winter," a concept of global climatic catastrophe brought on by even a "small" nuclear war, which could reasonably (the logic is compelling) lead to humanity going the way of the hapless, helpless dinosaurs. The key ingredient of this theory is the global climatic role of atmospheric dust, and without studies of dust storms on Mars and Mesozoic dinosauricidal asteroids, it is unlikely that the theory will ever be provable, except by a planetwide one-time experiment.

These conceptual impacts do not exhaust the continuing harvest of observing Earth—and other worlds—as planets, from space. Other equally valuable but equally unpredictable discoveries lie ahead as new sensors and perceptual technologies are brought to bear on the vast areas of global ignorance. More than mere curiosity is involved, although that may have been the cause of initial interest. To manage this world, to survive on it, we must understand it and the impacts we are making and might make on it. The view from space makes such capabilities possible.

chapter six

———

LIFE
CYCLES

But for the rivets, we would have thought it a scenario out of Greece's Golden Age, in which an oracle leads the neophytes deep into the cave whence springs the water of life. Our guide was a short, cheerful woman in her mid-thirties, eager to show off the fruits of her special art. We followed a twisting and turning path through the bowels of the structure, the taller of us occasionally ducking abruptly under mysterious pipes and air ducts. Our steps were circumscribed by bright yellow and black striped zones of peril—"Abandon hope all ye who step here," or some such dire warning.

Through a final door, we came upon the sought-for apparatus. Like a leftover from some mercifully long-forgotten Hollywood science-fiction extravaganza, the assemblage possessed an appropriate assortment of dials, lights, glass tubing, buttons, and switches. And, at the far right end, there was a 1930s Frankenstein-style beaker with water dripping into it from a pipe.

Our guide picked up the flask and proudly offered it to us. "It's distilled urine," the spacefarer trainee boasted. More soberly, admitted Dr. Mary Cleave (who earned a Ph.D. in sanitation engineering before being chosen as an astronaut in 1980), "We still haven't been able to adequately filter out the aftertaste of formaldehyde or the smell of iodine."

We sniffed the beaker cautiously. Admittedly, the water's

pedigree was not immediately obvious, but still, had this come from our kitchen faucet, a call to the county health inspector would have been the appropriate response. "It's certainly good enough for wash water," Dr. Cleave went on courageously, "and it should already be good enough for brewing tea or coffee."

We retraced our steps into the bright Texas sunlight, having seen a bit of the distant past and the near future. And we thought of the natural processes of Mother Earth: she daily and effortlessly performs the recycling miracles that space engineers are still struggling to master (constrained as they are to light, low-power equipment that has to work in weight-lessness). But to give them credit, Earth has had billions of years to learn how; the engineers have come pretty far in only a decade or two. A decade more should be plenty of time and, with the permanent American space station coming up, it will also be barely in time.

The massive, multifaceted biosphere of Earth provides its occupants with a sustaining environment, and we rarely notice the process unless it is inadequate. But what nature does automatically on a grand scale, people in space must do deliberately, on a small scale.

Beyond the limits of the biosphere, every hitherto unnoticed environmental condition must be purposely sculpted, monitored, and controlled. Conscious active efforts must replace unthinking passive acceptance. Like a nomad in the desert, a spacefarer is aware of water, since the fact of its absence is glaringly obvious; like a miner deep inside the Earth, a spacefarer is aware of breathable air, since the possibility of its absence is always near; like a nineteenth-century whaling captain far at sea, a spacefarer is aware of the inexorable day-by-day decrease of consumable supplies, since they are replaceable infrequently and at great effort. Only by thinking in these ways, which is rare on Earth, can people maintain a livable space habitat over many months.

Traveling in space, a voyager can through such efforts recreate many—but by no means all—aspects of the earthside environment. To the extent that it is possible to do so, artificial space environments maintain similar atmospheric composition, pressure, and temperature. But to the extent that the space environment cannot be made to mimic fully earthside conditions—most apparently in the conditions of weightlessness and radiation exposure—the human body suffers, partially adapts, and becomes subject to new forms of dying.

Sometimes the obvious must be made explicit. Air is the most urgent life support need. There must be sufficient oxygen, and there must not be too great quantities of harmful gases.

On Earth, the fifteen pounds per square inch (psi) sea-level pressure is created mainly by a mixture of oxygen and nitrogen. Oxygen alone contributes 20 percent, or three psi. In the lungs, this pressure is diluted by water vapor and carbon dioxide, so the actual pressure at which oxygen transfer into body tissues occurs is only about two psi.

Even lower pressures are tolerable, within limits. At an altitude of six thousand feet, lung oxygen pressure is 1.5 psi, and nearly complete acclimation is possible. At lower pressures (equivalent to higher altitudes), problems increase: although people can live continuously at an altitude of twelve thousand feet, the lung oxygen pressure of one psi requires extensive physiological adaptation, and physical performance is reduced.

In any case, an adequate supply of oxygen must be provided for a spacefarer. Dangerous gases, particularly the carbon dioxide that we exhale, must be removed. Buffer gases are optional, but if present must be maintained. During the reading of this paragraph, you took several breaths without thinking about the need to perform these functions. But you are not in a spacecraft.

For early, brief missions in the 1960s, fully expendable systems ("open systems") were entirely adequate. Oxygen was brought in high-pressure tanks. Water, too, was stored as a supply brought from Earth, and discarded (or stashed away in a garbage bag) after use.

Carbon dioxide was controlled by running the cabin air across chemical beds of lithium hydroxide (its chemical formula, LiOH, is commonly pronounced much like a refrain in a Western ballad about lovers' deceptions—"Oh, Lie-oh, lie-oh!"). The gas is removed by the process of adsorption: the carbon dioxide molecules stick to the exposed surface of the lithium hydroxide crystals. But the process soon slows down as all the crystal surfaces of a new cannister become covered.

On space shuttle missions, each seven-pound cannister is rated for about two man-days of usefulness. Up to thirty cannisters can be carried in the lower-deck closet, and two are hooked "on line" in the air system at any one time. In an emergency, if higher CO_2 levels were acceptable, "used" cannisters could provide some additional service.

In Earth's atmosphere, carbon dioxide is present at a concentration of about 0.060 psi. On a spacecraft, the allowable levels are somewhat higher: the shuttle's normal upper limit is 0.150 psi (in practice is several times less, near 0.060), the Salyut's is 0.170, and Skylab's was about the same. For emergency mode, a level of up to 0.250 psi could be tolerable for several weeks.

Safety requirements call for donning breathing masks if the concentration reaches 0.400 psi. Without such a precaution, adverse physical effects build up: by 0.750 psi, the pulse and respiration rates immediately start to climb; by 1.200 psi, the pulse reaches 90 beats per minute (versus a normal 70) and respiration more than 20 breaths per minute (versus a normal dozen or so). Headaches occur. At higher concentrations or longer exposures, the body's acid/base balance becomes seriously disturbed, leading to death.

A recycling CO_2 system was used on Skylab in 1973–74. There were two beds of silica gel: while one was adsorbing the unwanted gas molecules, the other—already nearly saturated—was being baked to drive off the CO_2 and dump it overboard through a small vent. When the first bed reached saturation, the second took over the air-purification duties while the first was purged. The system maintained safe levels, but it still left several times as much CO_2 in the air as did the expendable LiOH systems it replaced.

The Salyut space station uses an interesting process for replenishing its crews' breathing oxygen. Since it is difficult to build the complex equipment needed to regulate the gas flow from pressurized storage bottles reliable enough for years-long operation, and it is hard to resupply once exhausted, Russian engineers picked a deceptively simple chemical trick that might first have been described by Jules Verne. It weighs somewhat more than an optimal mechanical system but it is extremely reliable and very easy to resupply.

The Soviet spacefarers use cannisters of a chemical called potassium superoxide arranged in sheet form inside a 150-pound golf bag–sized unit, which is good enough for about six man-days of oxygen. As cabin air is circulated through the cannister, water serves as a catalyst (a material whose mere presence encourages a reaction to take place, but does not itself "get involved") to allow carbon dioxide to react on the surface, forming potassium carbonate (a solid residue) and free oxygen (which enters the cabin air).*

There is, of course, much more water in a spacecraft than just that dissolved in the air as humidity. Tanks of drinking and wash water are sent up full, either with the vehicle at launch or aboard supply ships. Water passes through the spacefarers and comes out as moist breath and urine; the former builds up and is controlled, while the latter is dumped overboard.

These technologies allow spacefarers to keep their cabins adequately supplied with breathing oxygen, while keeping

*For the chemists in the audience it goes like this:

$$4KO_2 + 2CO_2 \xrightarrow{2H2O} 2K_2CO_3 + 3O_2$$

their air clear of excessive levels of dangerous gases; they make available adequate water for drinking, washing, and food preparation. But these techniques require a tremendous amount of continually replenished supplies and equipment (on Skylab, for example, six tons of water were sent up when it was first launched, for a planned fifteen man-months of usage). For large numbers of people staying permanently in orbit, reliance on such time-tested technologies would require enormous freight shipments from Earth.

At one of the myriad farsighted private spaceflight symposia that have dotted the calendar in recent years, Phil Quattrone of NASA's Ames Research Lab outlined the basics of recycling life-support material. The formulation we heard was at Boulder, Colorado, in 1981; similar pitches were being delivered elsewhere and elsewhen. Quattrone specified two types of equipment to solve this problem: one is called an "air revitalization system," the other a "water reclamation system." Both are fully mechanical/chemical systems for processing waste gas and fluids into material usable in the place of new consumable supplies.

For example, some advanced techniques are now available for carbon dioxide processing. They include an electrochemical process, the Sabatier Process, and the Bosch Process. All of these involve the use of hydrogen gas, which is reacted in various ways with carbon dioxide to form water, from which the oxygen can subsequently be recovered. In practice, the first technique can transfer the carbon dioxide from the cabin air into a different reservoir, and another technique can then process the unwanted gas into other chemical forms.

The EDC Module (for "Electrochemical Depolarized CO_2 Concentrator") developed at Ames Research Center runs cabin air through a chemical module. Reactions within the module are fueled by hydrogen gas, and consume free carbon dioxide and some oxygen as well. The net result is that carbon dioxide is removed from the cabin air and concentrated in a sealed container; in addition, electrical power is given off, and can be shunted into an adjacent piece of recycling or storage apparatus.*

The payoff is that as with a stage magician's prestidigitation, the CO_2 ("Keep your eye on the molecule") has "passed through" an airtight barrier to a place where it will not harm the onboard breathers. Such units have already been tested at Ames, and are considered sufficiently mature for space-station use.

The Sabatier Process involves combining hydrogen and carbon dioxide by means of a high-temperature catalyst. The result is water, methane (CH_4) that is vented overboard, and heat. Some units show single-pass efficiencies of as high as 98–99 percent. While the device can be repaired only by a specialized electrochemist, the equipment can be designed for modular replacement of failed units. It's an extremely promising technique but has not yet been tested in space.

The Bosch Process uses a hot-iron catalyst to force carbon dioxide and hydrogen into solid carbon and water vapor. The process has a low efficiency (typically about 20 percent) so repeated passes are needed; carbon monoxide is created temporarily during the repeated cycles, which adds to the potential hazard. The carbon is deposited in replaceable cannisters that are discarded. This technique is not a leading contender for space stations of the coming decades, but the Sabatier Process's need for expendable supplies of hydrogen might make Bosch attractive for very long-lived systems.

Elementary chemistry students have all produced free oxygen from water by the simple process of placing electrodes into the water and using electricity to break the oxygen atoms free from the hydrogen atoms. The process is called "electrolysis," and on Earth depends on the weight-induced flow of the gas bubbles "up" to the surface of the water. Under weightless conditions of spaceflight, this process rapidly loses its simplicity, but American engineers have already developed two candidate systems for space station use.

*For chemists it goes like this:
$2CO_2 + O_2 + 2H_2 \rightarrow 2CO_2 + 2H_2O$ + electrical energy and heat

Nitrogen gas, too, must be generated aboard a large long-lived space habitat, to make up for steady leakage and for loss through airlock dumpings. Storing it in pressurized gaseous form is heavy, bulky, and potentially dangerous, so the most feasible plans call for storage in a chemically combined and easily handled form, such as hydrazine (N_2H_4). Relatively simple electromechanical systems exist to break the liquid hydrazine down into nitrogen and ammonia, and then break the ammonia down into more nitrogen plus free hydrogen. Under such a scheme, the hydrogen is then conveniently available for use in carbon dioxide–removal reactions, while the nitrogen is fed directly into the habitat's air.

One can have entirely appropriate levels of oxygen, carbon dioxide, and other gases, and still have an unbreathable atmosphere if it is contaminated. Watering eyes have struck Gemini space walkers and space shuttle crewmembers. The first crew to board Skylab proceeded with extreme caution due to fear of poisonous fumes that might have been driven off from materials during the station's unplanned overheating. One Salyut mission in 1976 may have been cut short by an "acrid odor," and another Soviet crew in 1983 was reportedly bothered by similar symptoms. Careful air measurements by cosmonauts showed how acetone and acetaldehyde levels soared during unloading operations from Progress freighter ships and during physical exercise. And trace contaminants that do not affect human health have been found to poison growing plants during long Salyut missions.

In a 1984 life sciences report on space station issues, Dr. Martin Coleman, a Ph.D. toxicologist at NASA's Johnson Space Center, wrote: "Considering the long period of time that atmospheric contaminants could accumulate and the great diversity of materials and equipment projected, the Space Station could present a greater toxicological hazard from atmospheric contaminants than was seen during any of the earlier orbital missions."

Controlling air contamination is crucial to the success of all space activities. Alarm systems, diagnostic equipment (such as a gas chromatograph/mass spectrometer), and eye/respiration protection devices are needed.

Most contaminants can be controlled by passing the air through filters of activated or chemically treated charcoal. But Phil Quattrone cautioned that "No single contaminant control process is suitable for all contaminants—some gases will poison oxidation catalysts and must be removed by pre-sorbent beds to protect the catalyst." Some gases when oxidized form extremely toxic substances, for example, harmless fluorocarbons convert to deadly carbonyl fluoride.

Sometimes the cabin air just smells bad. Early Gemini and Apollo crewmembers were reminded of a particularly pungent combination of locker rooms and outhouses (one slightly seasick frogman leaned into a just-returned Gemini capsule to shake hands with the astronauts, took a deep breath, and abruptly vomited into the commander's lap). Despite best efforts at personal hygiene, the problem remains that stomach gas cannot go "up" to get belched politely out the mouth; instead, it is processed out the other end of the gastrointestinal tract, "very effectively with great volume and frequency" (in the words of Skylab space-doctor Joe Kerwin). But after a while, a spacefarer tends no longer to notice the aura. The Soviets reported experimenting with ionizers and various fragrances (such as "pine forest"). Cosmonauts have not commented on the effectiveness of either.

Air must be kept moving, since, without convection, warm air doesn't "rise." Fans were supposed to keep most of the Salyut air moving at up to two miles per hour, but in practice averaged about a third of that speed. One place in particular showed nearly stagnant air: the bunks. Cosmonauts complained about feelings of suffocation when they got into their sleeping bags, but CO_2 measurements were normal. In 1983 Soviet engineers finally determined that the slow air movement was the culprit.

Once your air supply is safely ensured, your next concern—on Earth and in space—should be water. Bringing fresh water and throwing it away after a single use is tolerable only for relatively short missions, and in recent years Soviet space-farers have been recycling half of their supply.

To reclaim fresh water from waste water, there are several promising techniques. Quattrone labeled them "multi-filtration," "phase change," and "membrane processes."

For low concentrations of contaminants—say, cabin humidity condensate, or dilute wash water—a series of specialized filters (such as for particles, for bacteria, and for anion exchange) may be sufficient to restore the water to potability. It is in this area that the Soviets accumulated many man-years of experience in the early 1980s. While it is an "easy" step, it in fact allows half of all "waste water" to be recovered and reused.

Phase-change techniques occur naturally on Earth, as sea-water evaporates, condenses, and falls as rain. To mechanize this process in a manageable apparatus for a spacecraft, of course, takes considerable ingenuity, particularly to save and recycle the energy used to evaporate the water in the first place. Several distinct methods use this physical process, but most of them require various types of pretreatment and post-treatment of the water, using expendable chemicals.

Urine is understandably the most difficult to process. Quattrone deemed it "very complicated"—even with chemical pretreatment and posttreatment "you always wind up with ammonia somewhere." His favored procedure for this problem is "Vapor Phase Catalytic Ammonia Removal." The waste water (urine) is vaporized, mixed with an air stream, and fed into an oxidation reactor. Ammonia, urea, and light organic compounds get oxidized here, and water vapor is then condensed out of the steam. "The quality of the recovered water meets U.S. drinking water standards," he declared, "with the exception of a low pH. . . ." In any case, he continued boldly, "Ammonia is non-detectable in the condensed water."

American motivation to reprocess water (cabin humidity recovery is relatively easy, as the Soviets demonstrated) was mitigated by the overabundance of water on U.S. spacecraft. Gemini, Apollo, and space shuttle use "fuel cells" to generate electricity, and they do so by "burning" hydrogen and oxygen. Out come the watts, and a waste product—water! Properly filtered (and "debubbled," since some gas always seemed to remain in the water), the water provided more than the entire crew's requirements. Excess water on the shuttle is regularly dumped overboard with the urine, and this leads on occasion to problems such as the "Iceberg-1" satellite of 1984 (the hunk of ice that blocked a water-dump port and eventually had to be knocked free by the robot arm).

Storing water for long periods creates problems of its own, since potability can suffer from chemical and microorganism contamination. The Soviets solved this problem by going back about five hundred years to when Russian peasants left silver coins in cisterns and jugs to keep the water fresh. The physical explanation is that silver ions are adsorbed on the surfaces of bacterial cells, disrupting their metabolism and keeping them from multiplying enough to spoil the water. For years, cosmonauts have sworn by the taste and tonic of "silvered water."

Another urgent question of organic cycles in any environment is what to do with the waste products. On early, short space missions, feces were deposited directly into plastic bags, in which germicide was released to prevent gas buildup and bag rupture (the technique was at best "partially successful"). Vacuum drying (opening an overboard vent pipe from the sealed potty after use) worked well on Skylab, and, eventually, on space shuttle missions. On Salyut, the cosmonauts collect the material and stuff it into trash bags that are periodically dumped overboard, thence to rain down on the unsuspecting world. On really long-duration space habitats, the waste must be oxidized to liberate most of its water, but the process is a difficult one mechanically, and worthwhile only if the material is needed for growing gardens.

Quattrone has described a complete system utilizing these subunits, called ARX-1 (for Air Revitalization System—Experimental). It uses a nitrogen supply system with hydrazine that feeds both cabin air and the EDC module for CO_2 removal. The removed CO_2 is processed by a Sabatier reactor, which dumps methane overboard and passes the product water to a subsystem that combines it with extracted cabin humidity for processing by an oxygen generator. The unit, together with a control system designed for operation by a space station crewmember, is ready for flight testing aboard a shuttle mission and for installation aboard the space station.

Adding up all of the pieces gives a "map" of the mass balance of such a closed-loop system. Quattrone's system is designed for an eight-member crew. If everything is supplied fresh and used once, about 120 pounds of life-support supplies would be needed every day (fifteen pounds per spacefarer); if the recycling described by Quattrone was implemented, the total would be cut to only thirty pounds per day, a reduction of 75 percent.

In the closed-loop system designed at NASA-Ames, all water and oxygen are recycled. The only nonrenewable supply is food. Some chemicals and mechanical devices, such as filters, must be resupplied; hydrazine is also needed to maintain nitrogen levels.

Conveniently enough, nobody is ever going to have to drink his own urine. Drinking water comes 60 percent from recovered air humidity, 15 percent from the CO_2 removal reactor, and 25 percent from the CO_2 reduction reactor (the Sabatier processor). Of the urine produced in the habitat, 80 percent is used up in the oxygen-producing electrolysis reactor, 17 percent is excess water not needed anywhere in the life cycle, and the remaining 3 percent is lost as irrecoverable brine and is dumped overboard. None of it ever has to be drunk, or even washed in, even indirectly—barring spills!

In practice, it may be prohibitive to maintain two duplicate but separate water cycles. But even then, appropriate mental exercises in accounting can convince the squeamish that the urine is going somewhere else.

For an eight-person crew, twenty pounds of wet food per

day goes into the system. With wrapping and packaging, that comes to better than six tons per year. Hydrazine is consumed at a rate of about eleven pounds a day, making up for leakage of nitrogen (that requires two tons per year). Apart from less than a ton of supplies for the equipment, that is sufficient to keep the eight people alive indefinitely.

Without recycling, the team would need fourteen tons of water per year, and almost three tons of oxygen. They would need tons of lithium hydroxide cannisters to soak up the carbon dioxide.

The weight of the recycling equipment and the electricity-generation devices needed to power them comes to several tons. So there is obviously a break-even point, short of which it still is cheapest to go "fully expendable," but beyond which the initial cost of the recycling equipment is rapidly made up. It is "obvious" that a space station should have such a system, but that, alas, is not how space stations are designed.

Just how much of this recycling technology will be base-lined aboard the American space station remains an open issue. A trick of bookkeeping may frustrate the hopes of life-support engineers to install advanced systems and flight test them in the 1990s in preparation for use aboard manned interplanetary vehicles of the following decade.

When NASA promised the White House an $8 billion space station, it carefully delineated what that price would and would not cover. The hardware is part of the price, but the launch and operating costs are not. So in order to keep the front-end price down, advanced (and expensive) regeneration systems may be skipped over in favor of less-challenging tried and true technologies, even though this will clearly cost more in the long (or even relatively short) run. The "long-run" costs are not under the inflexible $8-billion roof, and by the end of the 1990s they could equal or exceed the initial costs.

"We haven't won this battle yet, even for the space station," warned Joe Sharp of NASA-Ames at a spaceflight conference in July 1984. Open-loop atmospheric systems were being recommended by headquarters officials, he disclosed, even though the operational costs would be "horrendous." They had the advantage that the up-front costs were much more "reasonable."

But NASA-Johnson budgetary official Humboldt Mandell reassured the same audience that "We will give up a lot of things before we give up closed loop." Mandell explicitly promised: "We *will* close water," but he cautioned that "other technologies will have to fight their way in based on economics."

At an earlier conference on manned flight to Mars, Phil Quattrone testified to the importance of space-station testing of the experimental technology. "We need [the space station]," he proclaimed. "If [it] doesn't go, we're not going to Mars." The closed-loop designer explained: "Given the go-ahead, we have to design the system eight to ten years ahead of time for the [space station]. Then we have to redesign it for Mars. So Mars is ten to fifteen years beyond the 1995 experience from the [space station]."

It is not a long conceptual leap from an artificial closed-loop regenerative life-support system in space to similar natural loops on Earth. Certainly the scale and number of variables are immensely larger, which is why the relationship between the concepts has not been widely recognized. And besides, the concept of a closed-loop spaceship life-support system goes back only a few generations.

But much of the recent earthside environmental ills can be traced directly to the conceptual fallacy of Earth in "open" or "infinite" terms rather than the spacefaring megahabitat it really is (Buckminster Fuller's "spaceship Earth" metaphor). Raw materials used to look "free," and waste materials could be safely dumped into a bottomless sink where they would disappear forever. Factors and influences could be divorced from consequences and results, and multiple mutually contradictory hypotheses about mass/energy flows could not be disproved. They were fit topics only for ivory-tower academics, irrelevant to the "real world."

That concept, of course, has been recognized as nonsense,

as the worst kind of head-in-the-sand cultivated ignorance and irresponsibility. As we learn to manage small-scale closed-loop life-support systems in artificial environments (in space, on the sea floor, or elsewhere), we will unavoidably be learning parallel and interrelated lessons about stewarding our natural habitat, on a planetary scale. The two concepts are opposite sides of the same coin, and the currency is life itself.

The space shuttle drifts top forward (posture unusual only by obsolete earthside standards) as it orbits Earth. On-board spacefarers come to view Earth and themselves from similarly unusual angles. (*Courtesy NASA.*)

Skylab orbits Earth, one solar panel lost and an emergency golden canopy unfurled to protect it from excessive heating. Only the presence of people allowed the multi-billion-dollar laboratory to function properly. (*Courtesy NASA.*)

Salyut-7, the Soviet space station, with add-on module attached, as it appeared in mid-1980's.

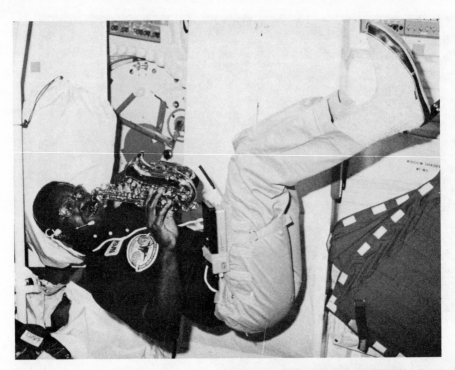

Spacefarers find brief diversion through music. Top, astronaut Ron McNair and his saxophone (*courtesy NASA*); bottom, cosmonauts Aleksandr Ivanchenkov and Vladimir Kovalyonok with guitar.

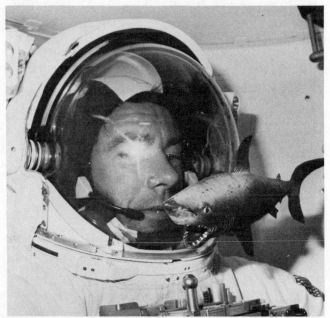

"Levity is appropriate in a dangerous business," said one space-farer. During preparation for a spacewalk, an astronaut found a rubber shark stashed in his suit by ground technicians. (*Courtesy NASA.*)

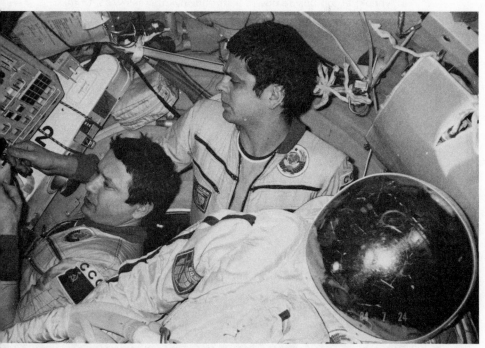

Temporary world space endurance holders (eight months) Leonid Kizim (left) and Vladimir Solovyov in airlock, preparing equipment for yet another spacewalk in mid-1984. (*Courtesy TASS-Novosti.*)

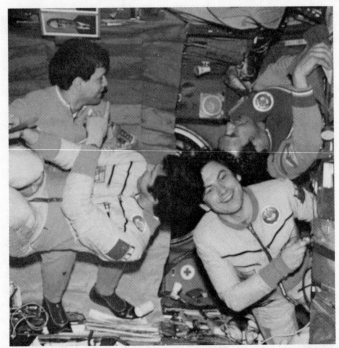

Soviet spacefarers celebrate the arrival of visitors to their space habitat. After months in space, they are accustomed to sitting on any convenient surface, having long ago abandoned earthside notions of "up" and "down." At bottom left, Vladimir Kovalyonok; at top upside down, Viktor Savinykh; cross-wise, Dumitru Prunariu from Romania; at bottom right, Leonid Popov.

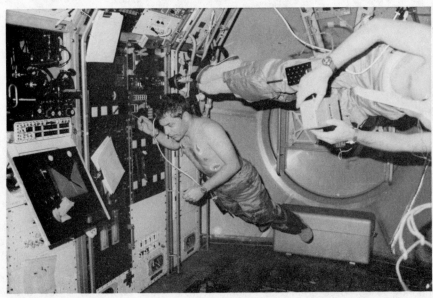

Everyday scene in space: American Spacelab scientists conducting research in orbit, disregarding obsolete notions of "floor," "walls," and "ceiling." Center, Robert Parker. (*Courtesy NASA.*)

Physiological effects of space travel include immediate major fluid redistribution in body. Note here how astronaut's face (Jack Lousma, in this case) becomes puffy and eyes assume almost oriental appearance. Subtle facial expressions can become unreadable. (*Courtesy NASA.*)

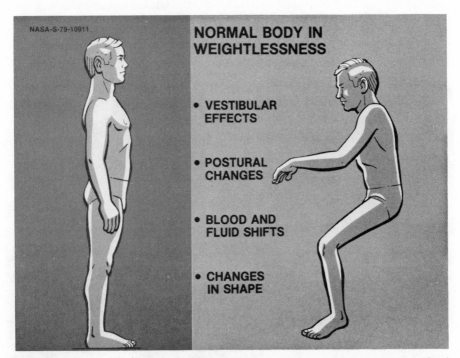

NASA-S-79-10911

NORMAL BODY IN WEIGHTLESSNESS

- VESTIBULAR EFFECTS

- POSTURAL CHANGES

- BLOOD AND FLUID SHIFTS

- CHANGES IN SHAPE

Under weightless conditions, the relaxed human posture assumes almost a fetal position (right); furniture aboard space stations must be able to accommodate this configuration. Earthside "body language" is made obsolete. (*Courtesy NASA-JSC.*)

"Ham" operator Owen Garriott chats with earthside radio amateurs during brief off-duty hours aboard Spacelab-1 in 1983. Bare-bones radio communication with Earth has proved inadequate for psychological health of long-term spacefarers, and must be supplemented by video links. (*Courtesy NASA*.)

Fanciful scene of spacefarer's video link to his earthside home underscores the importance of the space-to-Earth communications. (*Courtesy Motorola*.)

Shuttle mission crew patches demonstrate the class structure of spacefaring: professional astronauts on the "inner circle," with payload specialists and other guests in the outer darkness. (*Courtesy NASA.*)

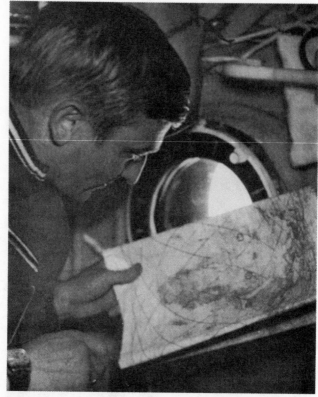

Spacefarers, both Soviet (top) and American (bottom, *courtesy NASA*), have viewed Earth through relatively small portholes on Salyut and Skylab space stations. Maps with station's orbital groundtrack tell spacefarers of both nations exactly where they are and where they're headed.

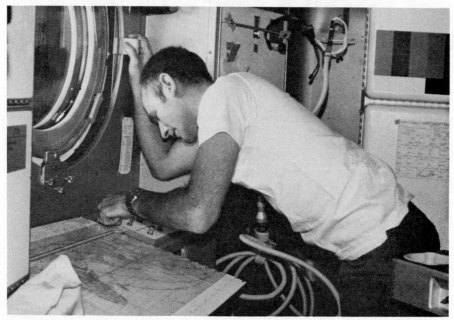

chapter seven

UNDER PRESSURE

M any delightfully awful science fiction or horror shows, from Dr. Who to James Bond, have the classic scene in which a hero or heroine is fed into a grinding machine that will turn him or her into puree. Consequently, spacefarers can be forgiven their trepidation when asked voluntarily to insert vital and beloved parts of their anatomy, generally below the waist, into a strikingly similar-appearing apparatus in orbit. Both American and Soviet space stations have been equipped with them. The doctors call them "Lower-Body Negative Pressure Devices." The spacefarers call them other things.

In 1984, the Soviets had cardiologist Oleg Atkov aboard one of their space stations, making a lot of special experiments feasible. One of these was the use of the lower-body negative pressure device, a pair of hooped trousers called CHIBIS. Because of his presence, Moscow spokesmen boasted that the pressure in the device could be lowered a full 35 mm of mercury, or 0.70 psi, below the cabin pressure of 15 psi. This would be dangerous without a doctor nearby, they claimed. Nobody had told the Skylab astronauts this more than a decade before. They had regularly run their pressures in steps down to a full 1 psi below the cabin pressure of 5 psi, with no ill effects.

The scientific purpose of such torture was to measure the

strength of the cardiovascular system by "pulling" blood to the legs, in mimicry of the effects of weight. Increased pulse and blood pressure could be measured as a function of "pull." The Soviets also found it a useful device for conditioning long-term crews prior to their return to heavy-weight conditions from orbit, where their hearts had "grown lazy."

Manipulation of air pressure in this device was thus a useful technique. On Earth we rarely pay attention to air pressure. In orbit, the choice of pressure, and the maintenance of that pressure, are crucial.

The issue of choice of cabin air pressure has not yet been properly addressed. The simplest answer is "Earth normal," 15 psi. The Soviets did it this way from the very beginning of their flights. They claimed, reasonably enough, that their common sense dictated introducing as few new environmental factors as possible so as to be able to isolate the physiological effects of uniquely spaceflight-related conditions. They may also have been influenced by a well-documented pure oxygen mockup spacecraft cabin fire that killed a cosmonaut trainee in October 1960. They refused to admit it, then or now, and, unwarned, NASA followed the same path, killing three astronauts almost the same way six years later. Eventually NASA too switched over to a 15 psi 20/80 mix of oxygen and nitrogen, and the issue seems settled for good.

But the question of air pressure has one last major impact on spacefarer activities. It is connected with the mechanics of space walking. The basic principles are straightforward. A certain minimum level of pressure must be maintained in the spacesuit, of course, to keep the spacewalker alive. This is 3.7 psi, to create the required oxygen pressure in the lungs ("normoxic" pressure); it needs a higher partial oxygen pressure when total air pressure is lower than normal.

Also, engineers prefer lower pressures, since this makes it easier to bend the spacesuit's joints, from the waist and arms and legs down to fine dexterity with the fingers. In general, bending a pressurized structure tends to decrease its volume, which raises its air pressure, which exerts a significant force against the bending motion. Additionally, means must be provided to bleed off the metabolic heat and exhaled moisture of the working spacewalker.

In practice, the problem arises not with the pressure of the main spacecraft (its governing factors have already been enumerated) or of the spacesuit itself (any practical value will do); it is in getting a human body safely from the former pressure to the latter pressure (getting back, if you survive, is no problem).

Since workers in the last century began construction and excavation activities underwater in pressurized compartments called caissons (or in individual underwater diving suits), a particular occupation hazard has been recognized— "caisson disease." It occurs when people "came up" to normal conditions after working at several times sea-level pressure. Excruciating pain often breaks out in the joints or the head; victims double up in agony (hence the term "the bends"), or lapse into comas, or die. By lengthening the transition from high pressure to normal pressure, such symptoms could be generally avoided.

The pains appeared to be caused by nitrogen gas that had been dissolved in body fluids under higher pressure, fizzing back into free gaseous form when the pressure was reduced— like a bottle of soda when the cap is popped off. In the body, these bubbles cause local pain, and in the bloodstream can cause blockages and consequent strokes (if in the brain) or heart attacks (if in the heart).

High-altitude fliers of half a century ago ran into the same problem on the other side of the pressure slope. This time, they started off at normal atmospheric pressure (at ground level) and faced nitrogen bubbles when they ventured into the much lower pressures of the thinner upper atmosphere. "The bends" could, and did, strike again, especially if pressurized suits or cockpits suddenly failed.

One remedy to this scourge was to find a way to purge the

body's dissolved nitrogen slowly and gently rather than abruptly and explosively. Breathing pure oxygen with a mask does this over a period of hours, so that any sudden decompression finds little remaining nitrogen to "boil off" inside the tissues and blood.

On space missions, the phenomenon is the same. It was confidently calculated, based on previous history—a century of human agony—that a drop from 15 psi (normal) to 8 psi was totally safe. But a further drop to 4 or 5 psi (the highest pressure at which engineers of the 1960s could design flexible-jointed spacesuits) would result in the bends. Prebreathing pure oxygen for five or six hours was sufficient to safely reduce nitrogen levels, but that added complexity and effort to the already overloaded early spacewalk schedules.

The American solution of the 1960s was to fill the spacecraft cabin with only pure oxygen, no nitrogen at all. Depressurization (either planned or unplanned) would no longer pose any problem, which was an important consideration. But in exchange for the simplicity and equipment savings the flammability danger was grossly amplified, and in January 1967 three men died in the Apollo-1 fire, during a pad test under the incredibly stupid condition of 15 psi pure oxygen (far worse than even flight conditions). That loss was easy to predict from the historical record, but determining whether deciding the other way would have resulted in an equivalent level of mortality from depressurization problems requires superhuman hindsight.

The Soviet engineering solution was to use an Earth-normal atmospheric pressure and composition with higher-pressure (and stiffer) spacesuits, combined with an acceptance of higher risk of the bends (keeping their spacewalks as short as possible also helped). That course, too, flirted with disaster. In 1965 their first spacewalker, Aleksey Leonov, would have died, unable to bend enough at the waist to get his legs back into the airlock, but for an emergency relief valve that let him in desperation dump most of the rest of the air from his ballooned suit. This consequently also posed the likelihood of imminent severe bends had he failed to get inside and repressurize—but not taking the chance was certain death outside, he later rationalized.

A decade of research by space medical specialists (one of them, Ellen Shulman, later became an astronaut herself, testifying to her confidence in the answers!) helped us to understand the phenomenon better, but effective precautions remain elusive.

Although diving activities can result in a wide range of symptoms leading up to choking, convulsions, and death, the smaller pressure differentials involved with space activities mean that much less severe results are involved, even under the worst conditions. Joint "awareness" and pain (mainly the knees and ankles, ranging from intermittent to incapacitating) are the most common expectations, although cerebral bubbles (causing such symptoms as headaches and vision problems) are conceivable.

Many factors seem to be connected with an individual's susceptibility to symptoms. These include the prebreathing protocol and pressure, amount and type of the spacewalking work, the spacewalker's metabolic rate (the higher, the more likely onset), and the distribution of body tissue types (more body fat leads to more trapped nitrogen, a physiological bias that predicates against women and that led prudent planners to add an extra hour of prebreathing to Kathy Sullivan's spacewalk preparation in October 1984).

If serious symptoms occur during a spacewalk, the crewmember returns to the airlock, and cycles back to full cabin pressure. Ground tests show that this relieves the majority of cases. Delayed symptoms following the walk can be treated by drinking fluids and breathing (via a mask) pure oxygen at 14.7 psi. The flight surgeon at Mission Control might also recommend steroids.

If needed, a modest "hyperbaric" (high-pressure) procedure has become available aboard American spacecraft: while in the airlock the still-suited spacefarer uses a special attachment fitting to overpressurize the suit with full cabin pressure all around, resulting in a suit pressure of about 50 percent higher than sea level. The Soviets may have already installed similar emergency equipment on their own space station, but they tend not to mention such possibilities.

In the unlikely event that such measures provide no relief, the crewmember must return to Earth immediately. For the space shuttle, this means that the whole spacecraft must land; aboard a space station, this requires special and complex evacuation techniques.

Interestingly enough, the occurrence of symptoms in no way precludes an individual making further spacewalks. Although physiological susceptibility seems to be repeatable for different individuals, a more thorough prebreath protocol for them could lessen the chance of future occurrence.

One other feature of spacewalking is dehydration. Body cooling is accomplished by webs of water hoses leading to a small refrigerating unit in the backpack, but under the high workloads typical of spacewalks, sweating continues. Together with exhaled moisture, this amounts to up to three pounds of water per hour. This must be removed from the suit's atmosphere lest it fog or freeze over on the visor (as happened to some early American and Russian spacewalkers), and the moisture must be replenished—a drinking hose is within reach by turning one's head.

Spacewalkers of the 1980s face the same physiological problems with somewhat better technology, but with a few new complications.

On the space shuttle, for example, American space engineers have finally adopted sea-level air-pressure standard. Prior to a planned spacewalk, the entire cabin pressure is dropped to 10 psi for at least overnight, providing some initial nitrogen purge from the crewmembers' bodies; the spacewalkers then do two more hours of pure-oxygen breathing in their suits before dropping the airlock to vacuum. Since most of the shuttle's electronics equipment is air-cooled, lower air pressure means less cooling and potentially higher operating temperatures and heat-induced failures. Calculations and inflight tests showed that the shuttle cabin's equipment could accommodate such a situation, but that the Spacelab module's equipment might not. So on Spacelab missions spacewalks must be preceded by the full four-hour prebreathe in the suits.

The temporary lower pressure means a higher partial pressure of oxygen, enough to raise again the spectre of inflight flammability. Obviously the 10.4 psi "step-down" is only a provisional solution, with problems of its own.

The American suit pressure is about 4.3 psi. The spacesuit itself has a separation plane around the waist (it breaks down into "shirt" and "pants"), with the upper torso a "hard" unit attached to the backpack. After one slips up into this unit, the "soft" lower section is pulled on and fastened with a waist ring assembly. Gloves and helmets are then put on and firmly attached.

The full American suit (and the Soviet one, too, approximately) weighs about 250 pounds on Earth, and nothing in orbit. But it still possesses "mass," which in practice means "resistance to change in motion," so extra effort is needed to start or stop moving around.

The Soviets still use higher pressure suits, around 5.8 psi, but their engineers have perfected the mechanics of joints and gloves to the point where mobility is adequate, even excellent. The Soviet spacewalkers use a hard-torso suit with integral helmet; the soft fabric legs and arms attach to this "cuirass." The backpack with life-support equipment is mounted on the back of the cuirass, and opens (in the practical words of one cosmonaut) "like a refrigerator door" to reveal the entry hatch. Like a molting insect in reverse, the spacewalker (already garbed in a garment covered with a network of coolant-water plastic hoses) slips his legs and upper body through the hatch, then uses levers on the front of the suit to close and seal the hatch behind him.

The Soviets boast of the flexibility of their suit: the gloved hands can pick up a coin off a table top, they say, and can use a ball-point pen; the arms are limber enough to touch hands behind one's back. There are no reasons to doubt these claims.

The suit's structural approach allows full access to the backpack's equipment when the suit is open and being serviced by the crewmember. In practice the Soviets have found this to be advantageous for maintenance, but it provides little protection between the equipment and the cosmonauts should something—a fire, a gas leak, an explosion—occur once the suit is sealed with a spacewalker inside.

At their higher operating suit pressure, the Soviet cosmonauts required only half an hour of prebreathing pure oxygen at 10.6 psi before the final pressure drop. This occurs in the airlock; the main station remains at 15 psi. Extensive ground tests, and extremely impressive orbital operations in 1983–84, demonstrated the sufficiency of this protocol.

For a space station, Soviet-style suits are the obvious necessary direction: one cannot partially depressurize a spacewalker's environment every time he or she is about to go outside. The low-pressure shuttle spacesuit is a technological dead end.

NASA-Ames has been working on this problem, and has designed a fully hard suit—a suit of genuine "space armor"— that operates at 8–9 psi and consequently needs no prebreathing at all. Advances in mechanical joints and seals allow the suit, including its gloves, to be even more flexible than the soft suits now in use. This is mainly the result of joint mechanisms that ensure "constant volume" (and hence no counterpushing air pressure) no matter what the joint position.

NASA's space station czar, John Hodge, made no excuses when he discussed the issue in late 1984. "We have to build a good glove. Working in those space suits for six hours is exhausting and this is especially true of the gloves."

Not only is the 8-psi suit and glove more usable, it is more maintainable. It can be stripped down and serviced on orbit, allowing dozens of spacewalks over many years of operation.

To turn around the current shuttle suit between missions takes 65 days and 80 people, according to Joe Sharp. For long-range missions, "Better send an army, or send a different spacesuit," he quipped.

Even as the space station program matured, the question of pressure had not been resolved. Early in 1984, a "Medical Sciences Space Station Working Group" in Houston threw another monkey wrench into the works. In a formal report covering many life science topics, the group came out against the 15 psi pressure for the space station. Herbert Greider, who had been designing life-support systems for NASA since Project Mercury, wrote: "The best solution is to select a space station atmosphere at an intermediate altitude, such as ½ sea level pressure, and use a suit pressure of approximately 4 psi." In other words, rather than redesign the spacesuit, NASA-Houston suggested redesigning the space station's life support!

The report "highly recommended" an air mix of 7.4 psi with 40 percent oxygen and 60 percent nitrogen. This would add problems, such as flammability—"[It] will require a somewhat more rigid material selection than a one-atmosphere cabin," it admitted—but that fell in somebody else's jurisdiction.

So there seems to be no conclusion to the continuing debate over pressure. Perhaps it is because this factor surrounds us so invisibly and unsensed. Maybe it doesn't seem "real" until someone dies in flames or in blood-fizzing agony. Then, to coin a pun, the pressure is on.

Many centuries ago, Marco Polo reported an extraordinary perception while crossing the Pamir Mountains in central Asia. He could stick his hand into boiling water without injury. While his readers back in Europe found this claim (as with a million others) incredible, in later centuries scientists realized he was accurately describing the effect of the low air pressure in those towering mountains in lowering the boiling point of water.

We can hardly claim to be any wiser than those who laughed

at Polo's original report. We, too, see the effects of air pressure all around us, as well as the effects of variations in it. Now that it has come time to venture into space, air pressure seems to be one of the last factors we can hope to get right.

chapter eight

GARDENS

It very well may be that historians of the far future will view the first establishments of self-sufficient, extraterrestrial communities of plants and animals (including humans) of Earth origin to be at least the second most important event in the long continuum of the living organisms of our planet. . . . For a biology of interplanetary and interstellar scales . . . the expansion of life from its planet of origin is much less common than is the origin of life on planets. This is why, at this larger scale, the relatively uncommon spread of species from their planets of origin may be more significant than the independent origin of life on different planets.—*Bassett Maguire, biologist, 1981*

Here all the processes in our organisms are sort of slowed down. For example, the growth of the nails on the hands and feet. But here in the "Oazis" in five days the stalks of dwarf wheat shot up 5–7 centimeters in length. It is pleasant to watch the greens; they are our nurslings.—*Aleksandr Aleksandrov diary, 1983.*

Anything that relieves the tedium, the boredom, the barrenness of life in deep space is in the interest of the Patrol. We have enough people going space-happy as it is. Flowers are considered good for psychotics on Earth; perhaps they will help to keep spacemen from going wacky.—*Robert Heinlein,* Space Cadet, *1948.*

Three-quarters of all households in America—and undoubtedly a like or greater fraction in Europe and Japan—have gardens. The plants range from decorative to recreational to dietary, from providing a pleasant ambience to providing the mainstay of the household's nutrition and livelihood.

While there are no plans for plants aboard America's conceptual space station of the early 1990s, the Soviets have been vigorously pursuing the concept of gardening aboard spacecraft. Many years of frustrating setbacks led in 1982 to a major breakthrough, the first seeds from a plant that itself had germinated in space. The cycle of plant growth had finally come full turn.

Growing plants in space may be of scientific and practical value, but the Soviet enthusiasm probably rests in large part on the simple fact that cosmonauts enjoy it. A major theme in Soviet first-person accounts of space-station duty is the psychological pleasure obtained from working with growing things. As early as 1975, during a sixty-three-day mission aboard the Salyut-4 station, Pyotr Klimuk remarked: "It pleased us to fuss around with the plants. With one glance at them our hearts grew lighter." But the men felt a corresponding disappointment when, after two or three weeks, the green things began to wilt and die.

Anatoliy Berezovoy spent seven months in space in 1982. He noticed that the garden had a soothing effect. "Recall how in moments of leisure, when we are sitting somewhere in the forest around the campfire, we can sometimes gaze at the fire without noticing the passage of time. And in this attractive contemplation we find calm. You experience something similar out in space as you look affectionately around your 'kitchen garden.' " In space, he continued, "you value their place in a new way. And we regarded the appearance of each little leaf, each shoot, as a victory in the struggle against weightlessness. Naturally, this was a source of joy."

But it was a fragile basis for joy, since for a long time the Soviet space botanists were unable to develop equipment capable of sustaining plant life and, most important, of growing a plant from seed to the point where it produced its own new generation of seeds. In 1979, a tulip in bloom was sent into space aboard an unmanned supply ship, but it died the next day. Lyakhov, the mission commander, recalled the event with anger during his postflight debriefing: "Judging by everything," he spat out, "nothing can survive in space."

A year later, specially selected orchids were sent into space. They arrived there in bloom, but on the station the petals almost immediately fell off. The experiment was repeated, and the petals fell off again. But the plants survived, and after return to Earth they flowered immediately. Why didn't they do so in space?

Valeriy Ryumin had been aboard during these setbacks, and he understandably developed a pessimistic attitude. "On Lenya Popov's birthday, an arabidopsis blossomed. . . . The plant was a tiny one and the little flowers were fragile and not very pretty, but we were happy! We guarded it and cherished it. To no purpose, however, because it turned out to be sterile."

Ryumin continued: "I was and am still somewhat pessimistic about this," he told an interviewer in 1982. "On both flights we took along seeds of the most variegated plants. There were those the biologists recommended and some that we cosmonauts carried on board almost as contraband. We planted them, everything was all right, and they grew. Before they matured, however, they withered away. And we tried everything we could think of!"

There was a vigorous debate between two main theories: one group insisted that there was some congenital condition of weightlessness (perhaps without convection, poisons and waste products accumulated in the organisms), while the other thought it was simply a matter of equipment and environment, which could yield to an engineering approach.

All sorts of devices and apparatus were developed. The "Oazis" unit evolved from flight to flight, developing into a minigreenhouse with lamps, ventilation, and watering systems, as well as a system for aerating the root zone of the plants (several layers of woven material perform the function of soil). Something called a "Bio-gravistat" involved a small centrifuge to produce pseudogravity; the early models were unsuccessful due to excessive aeration of roots. The "Svetoblok" ("light block") unit was a self-contained "space hothouse" first flight tested in 1982. And the "Magnito-gravistat" apparatus used magnetic fields to guide the growth of plants in the absence of geotropism.

A device called "Malakhit-2" was used on a six-month expedition in 1980. It had two lamps, four panels for plants, and an artificial ion exchange soil. Plants grew well, but they still refused to flower. This exasperated and frustrated the crew to such an extent that they resorted to a rather cruel practical joke on the biologists back on Earth. On July 30, they excitedly told via a television transmission that they had gotten a flower. They even showed it onscreen. The biologists were ecstatic and could hardly wait to get their hands on it when a visiting crew brought it back to Earth a few days later. Inside the container they found a perfectly crafted blossom—of paper!

(Upon their return to Earth, the cosmonauts were severely reprimanded for this little prank. Space medicine experts up to and including the chief, Oleg Gazenko, were not at all amused, and did not appreciate what looked like a sadistic and insensitive betrayal of the Earth-space link. The biologists, as it turned out, had been so trusting that in the few days between the televised image and the receipt of the counterfeit blossom they had excitedly prepared a series of revolutionary scientific articles, which they then had to cancel with considerable humiliation and embarrassment.)

The device that finally worked was "Phyton." It was developed by specialists from the Ukrainian Academy of Sciences' Institute of Molecular Biology and Genetics and from a botanical institute in Lithuania. Phyton included a special nutrient medium, as well as air filters to isolate the plants from potentially harmful impurities in the cabin atmosphere. More to the point, it also provided the plants with more artificial illumination.

The plant chosen for the experiment was arabidopsis, an unlikely candidate since it is commonly considered a weed that flourishes in quarries, dumps, and the semidesert. Noted one cosmonaut, "It would have remained obscure were it not for cosmonautics!"

Arabidopsis's chief attraction is its very brief life cycle, about forty days. Berezovoy and Lebedev planted seeds early in their 1982 mission, and a month and a half later reported success: the plants had flowered and produced seeds of their own. "The pods appeared," recalled Berezovoy. "Then they opened and we saw the seeds—there were about two hundred of them."

Some of the seeds were later planted on Earth, and most grew normally. Others were sent back into orbit on subsequent missions, and also did well. Soviet botanists justifiably considered this a major breakthrough. The cosmonauts were delighted, and put the new capability to practical use: when later in 1982 a visiting crew reached the Salyut, they (including Svetlana Savitskaya, Russia's second woman in space) were greeted by a tiny bouquet of arabidopsis flowers grown in space.

The cosmonauts tended more than just weeds in their orbital garden. All told, the 1982 mission grew ten kinds of plants: wheat, oats, peas, borage, radish, onion, parsley, coriander, dill, and carrots. None of these plants matured as well as the arabidopsis, but the cosmonauts were optimistic. "The biological experiments were not only of theoretical but also practical significance," Berezovoy noted, envisioning future vegetable gardens as a major source of food for space crews.

"Without plants, long-duration space missions are impossible," declared Lebedev. The two men had developed genuine affection for their growing things: "Before our return to Earth, it was simply sad to dig up the plants," Lebedev continued. "We took them out very carefully so that not a single rootlet would be damaged."

In 1983 another two-man crew (on a 150-day mission) concentrated on further experiments with arabidopsis and radishes. Shoots of tomatoes were grown for analysis, as were seeds of a promising grade of dwarf wheat. And the Ukrainian Central Botanical Garden provided an experiment for growing tropical epiphytic orchids, which, a spokesman pointed out, "can endure weightlessness and bring joy."

The shape of the radish was peculiar, the cosmonauts noted,

probably because it was grown under electro-stimulation. Vladimir Lyakhov, the crew commander, described it as "similar to a small tree." Later, his flight engineer confessed that "while we were packing it away Volodya and I broke off a few leaves. We tasted them and I for one liked the taste of that radish very much. In theory it could be used as a food in space right now."

One of the first groups of specialists to debrief the cosmonauts after their landing, according to Radio Moscow, were "biologists who are working on the program of creating hothouses in space, meant to supply long-stay crews with fruits and vegetables."

Lebedev had forecast space greenhouses in an interview a few months earlier. "These kinds of greenhouses will take up entire compartments in extraterrestrial stations, for plants need a different atmosphere from people, one with a high content of carbon dioxide and water vapor. The optimal temperature for obtaining the best harvest will also probably be different. . . . But the main thing is that they must have real sunlight." Such modules, Lebedev and other space officials broadly hinted, were already being constructed.

With the success of the space greenhouse, the positive psychological value soared. And beyond the mere pleasure of green, growing things, the gardens evoked—at a subconscious level—deep responses from the cosmonauts.

"Valentin Lebedev had never before grown plants," noted his crew commander, Berezovoy, in an article written a few months after their 1982 mission. "But on the station he used to rush off to the Oazis installation every morning almost before his eyes were opened. He was growing peas there. Valentin told me many times that it was on this mission that he first recognized the absolute fascination of this occupation."

Lebedev confirmed his shipmate's observation. "Back on Earth I had never loved tinkering in the garden. But on board the space station it was as if I woke up all of a sudden to the

Oazis apparatus. A tiny leaf opened up and it seemed to fling open a bright window out into the world. A pea pod stirs . . . and you're familiar with literally every little leaf. This was the first time I had felt a fascination in contact with plants. You sit down in your free time and watch them—they appear to be growing in front of your very eyes."

But in the case of this cosmonaut, perhaps the fascination with plants has a different genesis than a mere longing for Earth. Lebedev's wife Lyudmila was a member of the scientific team in Moscow that prepared the program of biological experiments, including the gardens. To be close to the plants was, for Lebedev, to be close to a project he identified—consciously or not—with his wife. But in any case it provided him with genuine psychological support under stressful conditions. "Future longterm space flights will have to include greenhouses," he told an interviewer shortly after his landing. "It is natural for man to busy himself with living things, and necessary for him to follow up his efforts."

The practical benefits of greenhouses would include better food, better health, better attitudes, and lower cargo launch requirements. At the turn of the century, Russian space visionary Konstantin Tsiolkovskiy had forecast it all. In 1912, he wrote: "In order to insure a food supply for a man during a flight in a spaceship, he must take with him various plants that will purify the air and bear fruit."

Exactly half a century later, only a year after the first manned space flights, the Soviet "Chief Designer of Space Vehicles" outlined a program of botanical research for space. Sergey Korolev initiated the research at the Siberian Branch of the Soviet Academy of Sciences in Krasnoyarsk that became the "Bios" program, but he died in 1966 just as the research was getting underway.

Several successful manned experiments took place at Krasnoyarsk, the longest lasting 180 days (a twelve-month three-man effort in 1967–68 was judged "unsuccessful," and such overambitious durations were not attempted again). The first

long Bios-3 run was a six-month stint in 1972–73; the biologists discovered that chlorella banks introduced toxic substances in the cabin air, which poisoned the food crops. In
1977, specialists Nikolay Bugreyev and Gennadiy Asinyarov
(termed "bionauts" by the Soviet press) made a months-long
run in which 60 percent of their diet was provided by food
crops.

The biological engineers noted that it was relatively
straightforward to construct closed systems at a level of 82
to 95 percent in terms of atmosphere and water. But repeated
attempts to push beyond this fraction of closure were unsuccessful due to what the scientists delicately termed "a
quite interesting problem."

The problem was connected with the use of mechanical
apparatus to "fully close the recycling loop." Such closure
was indeed possible, but only for the lifetime of the mechanical systems employed. When they had to be serviced,
they required more effort and more material than they had
saved in the first place. So the Soviet technicians fell back
from the push for perfect closure, and took stock of what two
decades of effort had accomplished. Their achievements had
been impressive.

Early in 1983 an article described an unnamed Russian
engineer who tested space equipment. He had recently worked
on spacesuit testing and wilderness survival exercises, but
his most interesting work had involved self-generating life-
support systems. At one experiment he reportedly spent a
month in an airtight room with a volume of only five cubic
meters (six foot ceiling, five feet square) "to test a system
employing chlorella algae to regenerate oxygen from the room's
carbon dioxide." According to the article, the CO_2 reactor
contained a thirty-liter suspension of chlorella. The tester's
food supply included algae crackers, a food other experi-
menters have found difficult to digest. The Russians must
have, as well: a subsequent experiment in a somewhat larger
room omitted the algae crackers and included higher food
plants such as wheat, carrots, beets, and lettuce.

Recent Bios-3 runs at Krasnoyarsk are by far the most impressive biological closed-cycle life support systems currently being tested. Early in 1984, two test subjects, Bios-3 veteran Bugreyev, and a new "shipmate," Sergey Alekseyev, completed five months of isolation inside a chamber outfitted with plants.

All purification of air and water was accomplished by plants; additionally, a special garden provided some food (the exact proportion was not disclosed). The room measured about thirty by forty feet, and contained three gardens (called "phytotrons") that covered half the floor space. A reporter described his impressions: "We walked up to one of the windows of the complex. On the other side of its glass it appeared as if it were a sunny summer day. Wheat on special tables was forming ears. In one place the wheat resembled a wall of yellow, at another place it was just beginning to mature, and at a third there were only green sprouts of seedlings."

This particular short-stalk wheat was specially developed at the Biophysics Institute. It has an impressive yield and gives six crops per year.

Dr. Boris G. Kovrov, deputy director of the Institute, boasted that "For everyday vegetables of which we selected more than ten types, we increased the yield by five to eight times without losing taste qualities." Not wanting to leave the potential application of this research ambiguous, the article concluded: "Spaceships and space stations with a closed organic-substance cycle are realistic," thanks to such work by the Krasnoyarsk Biophysics Institute.

One liter of chlorella in suspension gives up to fifty liters of oxygen per day. Already, such banks have been tested aboard Salyut stations, where it was discovered that the plants grow much faster than on Earth, by a factor of three to five times. Optimistic calculations showed that thirty to forty liters (eleven to fifteen gallons) of chlorella suspension should supply one man with sufficient water, air, and protein to sustain his life, which would be good news if humans could digest chlorella.

Other specialists see intermediate levels of botanical applications. "Small greenhouses will appear first of all to supplement supplies aboard space vehicles with fresh vitamins." One advertised use for the Phyton apparatus is to fill the cabin air with "life-giving plant scents and bactericides of a highly effective nature. . . ."

Related work in controlled-environment agriculture has applications to spaceflight in addition to earthside benefits. At the Institute of Experimental Botany in Byelorussia, Vladimir Soldatov and Denis Fedyunkin have reported developments in artificial soils for accelerated crop growth. "Whereas one square meter of ordinary ground soil yields one kilogram of garden radishes in 70 days, the same area of artificial soil can yield 10 kilograms in 21 days," they claimed. This technology has been applied to growing winter crops for Siberian cities, such as Norilsk, and aboard the new atomic icebreaker, the *Leonid Brezhnev*.

For food supplies fulfilling a major part of human dietary needs, the question of selection and mix becomes complicated. Which plants must be chosen, based on nutrition, palatability, relative yield, storability, and ease of replacement?

Nutritionists have broken possible space-garden plants into four groups: "protein accumulators"—beans, peas, soybeans, peanuts, wheat, and rice; "fat accumulators"—peanuts, rush nuts, and soybeans; "carbohydrate accumulators"—potatoes, wheat, sweet potatoes, sugar beets and table beets, carrots, cabbage, rutabagas, and kohlrabi; "vitamin accumulators"— parsley, dill, sweet pepper, onions, lettuce, sorrel, radishes, tomatoes, cress, spinach, and leaf cabbage.

American space biologists have not ignored this question, even though space gardens have no planned role for the 1990s NASA space station. Several important studies and symposia were conducted in the late 1970s. Research was undertaken at the University of Arizona's Environmental Research Laboratory in Tucson; in 1979, NASA-Ames sponsored a workshop on "Controlled Ecological Life Support Systems"; and in 1983 Mel Oleson at Boeing completed a major cost/benefit analysis of space gardens for American space vehicles of the next century.

The Boeing study considered a number of options: a simple system to provide a "salad bar" good for about 3 percent of the diet; a halfway system, providing 50 percent of the diet; and a near-total system, providing 97 percent of the diet, with the balance in the form of vitamins (such as B_{12}), seasonings, and miscellaneous condiments. They considered only a common, edible plants for which design data were available, a pessimistic limitation if Soviet claims of improved space-food crop production can be believed.

The salad bar idea wasn't worth the weight of the equipment, in terms of diet. However, Oleson pointed out perceptively, it "could provide psychological advantages—but it is not considered significant from a life support viewpoint."

No garden system proved worthwhile for flights to Mars and back. The missions simply did not last long enough to reach the breakeven point, considering the initial weight of equipment needed. But for manned space stations in the Earth-Moon system, the gardens—even under the pessimistic assumptions of contemporary crop yields—are clearly the way to go. For a 50 percent system in low Earth orbit, Oleson estimated reaching the payback point within six years; for the 97 percent system, it would take another two years before savings began to accumulate, but they would then do so at twice the rate of the halfway system. Similar times were obtained for a lunar base life-support system.

A leading American specialist on spacecraft ecologies is Dr. Bassett Maguire of the University of Texas at Austin. He frequently appears at space symposia, loaded with hard qualitative and quantitative specifications for such easily imagined orbital gardening. In designing a space-based semiclosed food chain, Maguire surveyed the nutritional roles of numerous plants and animals, as well as their input needs. He found that a reasonably workable system should supply "a combination of adequate quantities of soybeans, peanuts, wheat, and rice . . ." together with rabbits, goats, and/or termites ("because they can depend to a greater extent on plant parts which will be unusable by humans").

"This diet will provide for most nutritional needs of the people," he concluded. The exceptions include vitamin B_{12} (absent from a purely vegetarian diet), and "a few of the other vitamins, minerals, and amino acids will be marginal in quality."

Regarding these shortcomings, Maguire jokes that "in a closed loop system, a little cheating goes a long way." That is, once a system approaches 95 or even 98 percent closure, an inordinate effort and expense may be involved in reaching the 100 percent value. Even on the longest conceivable missions of coming decades—multiyear treks to the Asteroid Belt and beyond—it may be cheaper just to pack along a five-year expendable supply of these last few items.

The main garden, meanwhile, will need a great deal of attention and labor. Using standard estimates for physical arrangement, yield (not the Soviet enhanced yields), and usable crop fractions, Maguire produced an initial estimate of about 164 cubic meters (5,600 cubic feet) of hothouse per person. As a comparison, this is almost twice the volume of the Salyut space station, which normally hosts a three-man crew.

So an effort is needed to pack the garden into a more compact and efficient form. One way is to increase CO_2 levels to beyond the safety level for humans; another is to drop the level of oxygen, which results in "a useful increase in plant productivity because the rate of photorespiration of some kinds of plants will be reduced." This suggests that the garden module might best be sealed off atmospherically from the human module, and visits could be made using breathing masks and a simple double-zippered "airlock" chamber. And in fact this design is exactly what the Soviets have been talking about.

Whatever form the garden takes, the scope of the botanical processes can be estimated. Maguire detailed them.

Transpiration is the evaporation of water from the leaves of plants, and has the greatest impact on the flow of water through the organisms. It assists the movement of dissolved materials from the roots to the shoots and leaves, and is also important to cooling the plant.

Under different assumptions about optimum transpiration rates for the plants that comprise his complete space diet, Maguire estimated that an astonishing four to seven hundred pounds of water per day must be passed through the biomass that is dedicated to feeding each individual crewmember. During the same period, when a human is giving off about four pounds of water vapor, more than a hundred times as much is being given off by the plants that are to keep him or her alive!

And such transpiration, which occurs effortlessly on Earth, needs special artificial conditions in space. In the absence of weight-induced convective air flow, forced air flow must occur over the leaves whenever they are in light.

The plants need carbon dioxide, of course, and Maguire computed that the humans will produce far too little. Each person's crops must consume about five pounds of CO_2 per day, several times as much as he or she breathes out. The plants also need oxygen, about four pounds a day per crewmember, which is twice as much as the human needs for just breathing.

The plants must be lighted, either by natural or artificial light. If the former, there will be a great deal of waste heat to dispose of; if the latter, the waste heat will be less of a problem, but nonetheless significant. A complete analysis of all costs involved in both technologies led Maguire to conclude that sunlight concentrated by mirrors and filtered into the greenhouse through windows was several times less expensive than electrical lighting. But lights could stay on all of the time, or could mimic earthside diurnal cycles; in near-

Earth orbits, sunlight has a periodicity of an hour and a half. Could plants adapt to this?

Next, Maguire estimated his harvest: of the three pounds of dry weight of plant material needed per person per day, only 40 percent is edible. The residue consists of stems and other unusable parts. This is a greater "waste" fraction than the cans and wrappers that accompany prestored food into orbit (and that usually weigh about as much as the usable food they hold).

Maguire suggested waste eaters such as rabbits and goats because "they are small and have the advantage of minimum overlap of food requirements with humans—much of their food will be material which people cannot digest." Aside from the nutritional value of providing high-quality protein, vitamin B_{12}, and other nutrients in short supply in purely vegetarian diets, the animals would be of psychological value: "The inclusion of meat (and milk, yoghurt, and cheese?) will provide greater satiety value and gustatory variety."

Maguire particularly stressed the issue of microorganisms that would accompany gardens into space. They often have crucial roles in clearing away plant debris, and "sterile plants" just do not seem to grow. Yet unleashing a few hundred species of microorganisms in a space habitat without fully understanding their control systems on Earth is imprudent, to say the least: mutated forms could devastate the garden and might affect the crew as well. And human-hosted microflora can have similar deleterious effects on plants, Maguire cautioned.

Compared to these grandiose plans and forecasts, actual plant-growing experiments aboard the space shuttle are minor. Dr. Joe Cowles, a biologist at the University of Houston, has sponsored several seed-growing experiments aboard shuttle missions. Sunflowers and mung beans germinated well, but there were problems with proper orientation, since normal geotropic reflexes were not working.

Until longer-lived platforms are accessible (say, guest berths aboard Salyut garden modules), further advances for American space garden experiments are stymied. Yet the research is continuing in many countries. The payoff may be an increased understanding of botanical processes in general—surely it was a humbling experience for Boeing and NASA engineers to realize that they had to make numerous guess-timates and "SWAGs" (scientific wild-assed guesses) about crop performance for their study because the data simply weren't available. Studying crops and artificial means of raising them is part of a broader subject area called "Controlled Environment Agriculture" (CEA), and the earthside harvest of these studies, both literal and figurative, can be bountiful in years to come.

In the meantime, the realization is growing that when human beings enter space to stay, it must be not as loners but as part of a synergistic, symbiotic alliance of earthborn life forms. The gardens of space are showing the way.

chapter nine

SPACE DEATH, SPACE BIRTH

Death stalks its victim differently in space than on Earth. If a person went naked into the vacuum of space, death would be swift and easy. Air would be pulled out of the lungs with one quick exhalation, and blood going to the lungs from the circulatory system would also lose its oxygen. The shocking cold of space would hit him at the same time, and his skin might tingle unpleasantly as perspiration turned to ice. He would become giddy from the lack of oxygen, or might even try to laugh, although the sound could not carry and there would be no air to push the sound out beyond the vocal cords. The vision would blur and the person would pass out, with no real pain or discomfort, in about ten or fifteen seconds. It would take only a few minutes for the brain to die. The cause of death would be listed on a death certificate as "hypoxia," death from lack of oxygen.

After death, the body's water would seek to evaporate because of the absence of atmospheric pressure, causing the skin to blister all over. Some blood vessels would break, causing bruises here and there. Contrary to popular belief, the person's unprotected body would not explode, because the skin is fairly tough. It would take a few days for the water in the body to escape through the skin via evaporation. The body would thus be dried, mummified, as it were; it would not decay. It wouldn't even shrink much because the skeletal structure would still be intact, but it would weigh a great deal less from the loss of water.

If the skin were exposed to light, it would dry out and deteriorate, but probably wouldn't blacken or become tanned because the organic processes for that would have been stopped. Although the skin's resilience deteriorates over time, it would not break off from the body unless some mechanical force were exerted on it, a collision with some space debris or another object.

If this death occurred near the poles of the Moon or Mars, or in deep space beyond the Asteroid Belt, water would not evaporate out of the body, but would rather freeze solid. Again, the body would not decay.

Assuming death were accidental—on a spacewalk—say as a result of an accidentally ripped spacesuit or smashed helmet—the process of water evaporation and mummification might be slowed down somewhat. The dying process might take longer than fifteen seconds depending on how long it took the suit to depressurize and lose its vital gases. The person would still not have much time before losing consciousness to comprehend what was happening.

On a somewhat larger scale, a whole crew could die of hypoxia if a spacecraft sprang a leak and atmospheric pressures fell rapidly. People on Mars or the Moon might die of hypoxia if their life-support systems failed to deliver oxygen—say, due to the rupture of the oxygen tanks—but still managed to scrub carbon dioxide from the habitat atmosphere. Resupply of oxygen from Earth or extraction of oxygen from the Martian atmosphere would take too long. As their brains starved for oxygen, they might die laughing, as if it were all a big cosmic joke.

This is not mere theorizing, since men have already died in space exactly this way. In June 1971, on their way home from Salyut-1, the world's first space station, three cosmonauts lost their lives when an air-pressure equalization valve in their Soyuz-11 command module popped open, dumping their cabin air out into space within a minute. Georgy Dobrovolskiy, Vadim Volkov, and Viktor Patsayev died in their seats, and were found by shocked ground crews after a perfect automatic landing. Their faces were peaceful, although one had a large bruise on his cheek that was still clearly visible days later when they lay in state in Moscow. There were many wild speculations about the cause of death, including unknown medical effects of their record-breaking twenty-four days in space, but the precise answer came out a few years later as Soviet officials opened their account books in preparation for the Apollo-Soyuz orbital handshake. A simple mechanical flaw had led to hypoxia and death.

A more terrible form of extraterrestrial death is from exposure to large doses of radiation. On Earth, we are protected from such solar events by the magnetic field and the atmosphere, but there is no such safety in space or on the other planets. A large solar flare could kill. Accidents involving nuclear power—nuclear-powered engines aboard a spacecraft or space station, or a nuclear reactor on the Moon or Mars—might also occur.

If a person were exposed to a tremendous dose of radiation, ten to thirty thousand rads, during a nuclear incident, he or she would die within two hours from central nervous system damage. Within thirty minutes to an hour, he or she would collapse and go blind. Seizures and convulsions would follow, then unavoidable coma and death. The fictional alien Mr. Spock in The Wrath of Khan suffered exactly this fate; his death was accurately portrayed by Leonard Nimoy.

With doses of one thousand to seven thousand rads (the strongest solar flare ever recorded put out one thousand rads within seven days) a person might die quickly of central nervous system collapse, or might linger a few days. In that time, there would be much suffering. The person would experience terrible nausea, vomiting, and diarrhea as the gastrointestinal tract began to collapse. Death would be a welcome mercy within four to six days.

Solar flares putting out radiation levels of 350–750 rads would in the upper ranges almost certainly cause death from gastrointestinal collapse. In the lower ranges, people who didn't die within a month, of dehydration and emaciation, might do so from blood problems. Because radiation most strongly affects DNA, the body systems dependent on quickly reproduced cells—the gastrointestinal tract and the blood—are affected first. In the case of the blood, the bone marrow might be wiped out, resulting in loss of the clotting factors, the platelets, and the white blood cells that fight disease.

People exposed to 350 to 550 rads would experience not only nausea and vomiting, but would have fever and hemorrhage. The blood count would plummet, the platelets and white blood cells would be killed, and the person would experience anemia, uncontrolled bleeding from the nose, mouth, rectum, or skin lesions. He or she would have no resistance to disease. It would take a bone-marrow transplant to save this person, provided that secondary infections hadn't already finished him or her off. Less important but still unpleasant, the hair would fall out and there would be radiation burns on the skin. Those who recovered—and there wouldn't be many—would take at least six months to do so.

Dosages of 200–350 rads could kill 20 percent of a crew in two to six weeks; survivors might take three months to recover. People exposed to smaller doses, 100–200 rads, would survive, but with reduced blood counts and more vulnerability to illness. Only those with exposure to less than 50 rads would have no obvious effects.

Even if the individual survived, or did not absorb enough radiation to suffer these effects, the danger would not be over. Leukemia, skin cancer, thyroid cancer, breast cancer, sterility, and chromosomal damage might show up five, ten, or twenty years later. All types of radiation have a cumulative effect, so even constant low doses of radiation pose a threat over a period of time. A person who picked up ten rads a year for twenty years would have received the maximum lifetime radiation dose allowed to nuclear workers under current OSHA standards. To prevent space settlers from developing cancers, habitats on the Moon and other planets, and even space stations beyond the Van Allen belts, would have to be covered with heavy shielding and be provided with a massively dense storm cellar for solar-flare protection. Long-term habitats on the Moon and Mars should be covered by at least ten or twenty feet of dirt.

Finally, people in space will always be completely dependent on their artificial life-support systems that provide air circulation, the replenishment of oxygen, and the removal of carbon dioxide. A mechanical breakdown of the air-revitalization system could lead to carbon dioxide narcosis, another terrible death. In a large space station, it might take days for carbon dioxide to reach lethal levels; in a tiny spacecraft, it might take only a few hours.

On Earth, only a trace of carbon dioxide exists in the air we breathe. If the CO_2 in a spacecraft or space-station module goes up to 2 percent of the cabin atmosphere, people would notice. The cabin would become stuffy, and they would begin to breathe more rapidly. At 3 percent carbon dioxide, the breathing rate would be doubled, and people might have difficulty hearing. As carbon dioxide levels continued to rise, people would experience headaches, dizziness, nausea, mental depression, and blurred vision as they panted for breath. By 6 percent carbon dioxide, the crewmembers would become mentally confused, unable to take measures to preserve their lives, and would lapse into unconsciousness. As levels reached 8 percent, those who were still conscious would experience convulsions. The lungs would cease to work. People can tolerate such high carbon dioxide levels for only about ten minutes before becoming unconscious.

Of course, there are other, more familiar forms that death might take in space. The all-important heaters could fail, and spacefarers could die of hypothermia, since the natural average temperature of objects even in full sunlight this far from the Sun is far below freezing (our planet is warmer thanks to the misnamed but still effective greenhouse effect of the atmosphere). There will be a number of toxic chemicals aboard space habitats, and careless workers might perish from inhaling the gases. Fires may occur in which people die from smoke inhalation or burns. Spacecraft can crash, space workers can suffer penetrating wounds, people can have heart attacks, aneurisms, and strokes.

Medical treatment has its peculiar difficulties in space. If someone has a heart attack on Earth, cardiopulmonary resuscitation (CPR) is the first thing a rescuer does to get the heart started again. CPR can't be done in space; the victim and the rescuer both float, and it's almost impossible to exert enough pressure on the chest to pump the heart. In a hospital emergency room, intravenous lines (IVs) can pump heart stimulants or antiarrhythmics into the bloodstream. In space, that can't be done either, since IVs depend on gravity to drip the medicine into the veins. Pressure-fed systems require the insertion of a needle, which under weightlessness is so difficult that paramedic Robert Beattie of Wichita, Kansas, has suggested that all spacefarers have a venal shunt implanted below the skin of the inner, upper thigh. If the heart damage is very severe, requiring bypass surgery, the patient would have to be evacuated back to Earth or go without.

The good news is that some treatments of heart attack—balloon angioplasty, streptokinase injection—are unaffected by the space environment. Also, patients who survive heart attacks must rest, and space is the perfect place for that. Much of the heart's work is pumping against weight. Under ordinary space circumstances, the heart atrophies from lack of work to do, so a heart patient would not have any trouble resting there.

There are some clever ways to sidestep many of the major problems. The heart can be mechanically pumped by a thumper, a piece of equipment found in some earthside emergency rooms. The victim lies on a board while a gas-powered plunger thumps the chest. Its major drawback for space is that it is heavy, and the incidence of heart attack—currently zero—doesn't justify taking it along. "A more likely candidate is the CPR vest, which squeezes the chest like a blood-pressure cuff, and can by rhythmically inflated and deflated," said Dr. Daniel Woodard, an aerospace medicine resident at Wayne State University. "It's light, compact, and easy to use."

Space IVs will have to be specially designed. "They have to come equipped with a pump and a measuring device to monitor the infusion rate in a dripless environment," said Woodard.

The treatment of burns has its good news/bad news aspects as well. The good news is that fire as we know it, with flames shooting up, may not be able to burn because there are no convective air currents to keep it alive. To extinguish a fire one might need only shut off the fans. The bad news is that chemical and electrical burns are still possible, if not probable. Typically, badly burned people will feel faint, have a rapid pulse, have trouble maintaining their normal body temperature, and go into shock as a result of fluid loss. On Earth, treatment begins with covering the burn to keep body fluids in, and replacing the fluids that have been lost by means of a saline IV. At a burn center, dead skin is removed, and the wound washed with Clorox and dressed with antibiotic cream. Burn victims are always given large amounts of fluid and a great deal of food. Finally, more of the dead skin is removed, and a skin graft can be attempted, using the patient's own skin, cadaver skin, or artificial skin seeded with the patient's skin cells.

All these steps of the treatment have problems in space. The Clorox bath and the removal of the dead tissue would have to be done in some kind of enclosure since fluids do not pour and small particles float around. "Cleaning off the dead skin, or 'debriding' as it's called, requires some special equipment, but more important, very specialized training," said Woodard. It's not the type of procedure a general physician could be talked through over a radio link, since it requires a special training and a practiced touch. It's unlikely that a plastic surgeon skilled in debriding and skin grafting would be sent into space since general expertise is more valuable there. Skin grafting requires surgery under general anesthesia, and drugs of all types have somewhat surprising effects on some individuals in space. According to Dr. Bruce Houtchens of the Department of Surgery at the University of Texas, "In most hospitals on Earth, care of a single major burn patient outside a burn unit on a regular surgical ward imposes a major strain on supply lines, nurses, maintenance personnel as regards stocking clean dressing, performing frequent dressing changes, and disposing of contaminated dressings. These problems almost certainly would be multiplied several fold in the space station environment."

The problem of infection is probably no worse in space than on Earth and, with good hygienic procedures and limited access to the patient, may be somewhat better. And there would be no bed sores to contend with, no breathing or circulatory problems related to long bed rest. There is no weight to put pressure on the wound and to interfere with circulation and fluid drainage.

A large number of toxic chemicals are used on space vehicles, such as nitrogen tetroxide, freon, and ammonia. Exposure to them would be rare, but life-threatening. For instance, hydrazine is a very toxic chemical most spacecraft will carry as fuel, as a power source, or as a nitrogen reservoir. Inhaled hydrazine can cause death by convulsions or from pulmonary edema. Skin contact with hydrazine causes nasty chemical burns; the skin absorbs the poison as well.

On Earth the treatment for inhaling toxic fumes is to give the victim oxygen, IV fluids to prevent shock, anticonvulsants like Valium or phenobarbital, and steroids to reduce inflamation of the air sacs of the lungs. In space, the same could be done, provided one had the redesigned IV equipment, and doctors were sure that anticonvulsants and steroids would work in space as they do on Earth. External exposure is more problematical, since the treatment requires flushing the exposed area with large amounts of water. "It can be done," said Woodard, "by sticking the person in the shower, but you would have to make certain that the water would not get recycled like other bathing water and get into the ship's water supply."

The types of injuries that would strain space resources the most are precisely those which are most likely to occur: trauma as a result of a crash, from a rover, a transport, a spaceship, or even a long fall off a cliff. Assuming the victim did not die immediately from injuries or from hypoxia as a result of a ruptured space capsule or spacesuit, the prognosis still would not be good. These trauma are like car crashes on Earth: there are multiple fractures, sometimes ruptures of organs, massive bleeding, and quite often head injury.

The first strain would be on diagnostic equipment. CAT scans, NMR (nuclear magnetic resonance), even simple X-rays all use large, bulky, heavy pieces of equipment. And in space, it may be difficult to position the patient properly and get him or her to remain absolutely still while the picture is being taken. Houtchens points out, however, that portable X-ray machines existed decades ago when medical care was less centralized, more in the hands of the house-call doctor. Sophisticated miniaturized X-ray machines could be developed and manufactured if there were a need.

CAT- and NMR-scanners are huge pieces of equipment and unlikely candidates for space use soon unless some bright high-school kid does for CAT-scanners what Osborne did for computers: makes them small, light, and portable. Woodard pointed out that digital radiography would be a good candidate for space use since the patient can be rotated in the field of view, and the resulting picture could be similar to a CAT scan.

Then comes surgery. A sterile field would have to be maintained over a large area. Woodard suggested that one-way laminar air flow could direct contamination in one direction. "Or we could create a three dimensional sterile field," he said, "by attaching a large clear plastic tent. The outer surface is glued flush against the patient's skin, covering the entire area where the incision is to be made. The surgeon inserts his hand into built-in gloves and uses the instruments inside the bag. Any microorganisms on the patient's skin are immobilized by the adhesive." The Army thought of using this to treat field wounds a long time ago, but the surgeons preferred to operate indoors.

The surgeons might want to be rigidly fixed in position by feet and even waist restraints, with the operating table movable in relation to them. Special equipment of all types—clamps, retractors, and so on—would have to be developed since tissues, organs, blood vessels, and everything else would float slightly once the chest or abdomen were opened. Breathing and circulation might have to be supported artificially during the surgery.

Blood transfusions would be a problem, Houtchens pointed out, since the shelf life of blood is only about twenty-one days. Of course, crewmembers are always potential blood donors, assuming they too haven't lost blood in a serious accident. Fluorocarbon artificial blood is a possibility. A medical lab with various automatic and rapid blood analyzers has been the subject of study, not only for use in space but for remote and isolated medical treatment areas as well.

Most devices that can monitor blood gases before and during surgery are invasive; one, which Houtchens called a "transcutaneous oximetry and capnimetry device," monitors oxygen level and cardiac output through the skin. It is somewhat less accurate for adults than the premature infants it is used on now, presumably because the main blood flow in a newborn lies close to the skin's surface. Naturally, in space it would have to be made more accurate, or be combined with other methods of determining blood gases and blood flow.

A head injury would require drilling a hole in the cranium

and draining out blood and the other debris around the wound, calling for the extreme training of a neurosurgeon. Neurosurgery, like plastic surgery for skin grafting, is one of those specialties we are unlikely to find in space unless there is a large population and sufficient business to justify them. And, like debriding, brain surgery isn't the type of procedure one can be talked through easily by remote control.

Houtchens nevertheless recommends setting up a network of medical specialists on call on Earth to talk a space medic or general surgeon through delicate procedures. They could also form a "decision tree" to help diagnose (with good data uplink and clear two-way television pictures) and determine whether to treat a person in space with the limited resources or medevac the patient back to Earth.

Evacuation from orbit is no small feat. Most traumas call for instant treatment; they cannot wait a few days for a shuttle to fetch the injured person, even assuming there were one sitting on the pad already set for launch. A shuttle launch costs between $100 and $200 million, and each shuttle orbiter must go through rigid safety checks before it can fly. The shuttle doth not a good ambulance make.

Another method of medical evacuation has been proposed. It is called MOSES (Module Orbital Space Escape System), and it would be used, as Houtchens put it, to "evacuate a casualty to the surface of the Earth, possibly accompanied by an attendant." The module, to be located on a space station as an evacuation pod, is designed along the old Gemini lines with a capsule ocean splashdown and retrieval like the good old days.

Houtchens criticized the system, however, as "unlikely" for several reasons: most traumas need to be treated sooner than the two and one-half hours (from low Earth orbit) or six to eight hours (from geosynchronous orbit) it takes to get back to Earth; there is potential for a reentry and recovery mishap; if a doctor accompanies the patient on the module the station will be without medical care; and so on. "Most physicians will be unwilling to refer a sick patient to the Indian Ocean if there is any way they can maintain control," he stated. Evacuation from space, he concludes, is a poor alternative to providing adequate diagnostic and treatment facilities on the space station. Evacuation would work only for burn patients who have to wait for skin-grafting operations anyway.

Even if all life-support equipment works flawlessly, no industrial or occupational accidents occur, and there are no suicides, homicides, or executions, spacefarers are not out of the woods. In their attempt to adapt automatically to weightless space, the heart, the muscles, the blood, the bones, and the immune system all degrade in ways that can make eventual return to Earth a life-threatening proposition.

Because the aim of space medicine is to keep the human body fit for life in two regimes, space and Earth, space doctors insist on diet and exercise programs that are as tedious as they are necessary. Spacefarers have to quaff a good deal of water and stuff down at least 3,500 calories a day to keep from dehydrating and losing weight. The extra water seems to keep blood volume losses within the space norm of about 10 percent. The Russian doctors insisted until recently on roughly three hours of exercise on several types of devices—treadmills, bicycle ergometers, etc.—to keep the heart and muscles from atrophying to a dangerous degree. But many spacefarers hate to exercise in space.

"There is a real lack of desire to exercise," spaceflight veteran Valeriy Ryumin wrote in his memoirs. "On the ground, it was a pleasure, but here we had to force ourselves to do it. Besides being simple hard work, it was also boring and monotonous. . . . After about 10–15 minutes of workout, our faces and exposed parts of the body became covered with drops of perspiration. In space, drops of perspiration do not drip off the body, but take on a spherical shape without coalescing. Those water 'peas' are collected with a towel. At the end of the exercises, the sportswear was completely saturated with perspiration and had to be dried on the ventilators."

American shuttle astronaut Don Peterson felt the same way: "You and your clothes are soaked with sweat after even a short jog on the treadmill. You use three towels just to wash off afterward—a wet one, a dry one and one to wipe the soap off with."

Besides the potential atrophy of the heart and other body muscles, Russian flight surgeon Anatoly Yegorov in 1978 reported some changes to the immune system. The changes—a rise in steroid hormones and damage to the T-lymphocytes—could indicate prolonged stress or a breakdown in the body's ability to fight disease. So far, no illness has occurred as a result of the slightly lowered immune defenses.

Nonetheless, the Soviets thought it useful to send along physician-cosmonaut Oleg Atkov on their 237-day flight in 1984. He took blood samples from the veins and fingers in order to monitor hormone levels, carbohydrate and water-salt metabolism, and the condition of the blood at various points in the long flight. He evaluated some of the old equipment aboard, and oversaw a new exercise regime that called for a shorter but more intense and exertive use of an exercycle and running track. As a professional spaceflight cardiologist, he observed their hearts with an "Echography" device that utilized ultrasound to show heart action in real time. He performed numerous ear, nose, throat, and eye examinations, studying the dynamics of blood supply to these areas during adaptation to space. Last but not least, Atkov had psychological training as well, and was asked to study the psychological problems of small groups in confinement.

Atkov's crewmates Leonid Kizim and Vladimir Solovyov quipped: "We have to give the doctor what he wants." During a television session from space Atkov complained of being "the most unemployed doctor. My patients do not complain about anything."

An infection, though, might be reason for complaint. A joint French-Russian biomedical experiment called CYTOS suggested that there was what Pierre Langereux in an article for

La Recherche called "an appreciable decrease in the effectiveness of antibiotics under weightless conditions." Researchers noted that cell membranes of intestinal microflora bacteria became thicker and less permeable. Since antibiotics are the cornerstone of modern medical treatment, their in-space efficacy is likely to become the focus of some space medical research in the future.

All the effects to the heart, muscles, blood, and immune system are fully reversed once back on Earth. But some damage to the bones is permanent, and it's the bones that worry space doctors the most. The bones begin to lose calcium from the start of spaceflight. So far, there is nothing to suggest that such decalcification is a self-limiting process, that is, one that stops on its own once it reaches a certain level. During Skylab, the astronauts lost 7.9 percent of their bone calcium in three months. The Russians noted that calcium loss varies. The average is 14 percent over six months, although one cosmonaut lost 19 percent during a 140-day flight while others lost only 8 percent on a 175-day flight. During the 211-day flight in 1982 each of the two men appeared to have lost approximately 15 percent.

Exercise and supplemental vitamins and minerals have so far not prevented mineral loss from the bones. Not only do spacefarers run a greater risk of broken bones on their return to Earth, but they incur a slightly higher risk of kidney stones, especially if water intake had been limited or restricted during the mission.

Bed-rest studies on monkeys that simulate conditions in space have confirmed that it takes a long time to recover bone lost over six and one-half months of weightlessness. According to Dr. Don Young, manager of calcium research at NASA-Ames, there is "some new bone forming" three months after returning to normal conditions. At six months, there was some accelerated growth of certain kinds of bone, and after two years, there was still remodeling of other bones. Young noted that there was good restoration of cortical bone, the hard, exterior parts of bones, but not in trabecular bone, the inner, spongy-looking portions.

Under normal Earth conditions, bone is constantly being broken down and replenished with new bone. However, in space, the processes of breaking down bone are accelerated and the laying down of new bone interrupted. "The changes in the bones are dramatic and large," concluded Young, "and there is a question of whether bone loss is ever reversible. If you lose trabecular bone, there may be no way to restore it."

Some researchers have suggested that the problem is dietary, that calcium is not being digested and absorbed properly. Others look to the problem as purely one of physical loading of the bones. Suggested solutions include adding more vitamin D to astronauts' diets (vitamin D is essential for absorption of calcium) or even having astronauts sit under special fluorescent lights that mimic the exact spectrum of sunlight to stimulate vitamin D production in the skin.

Others believe that the answer lies in putting higher stress loads on the bones to keep them fit. Treadmills have never proven adequate to this task, but, according to Sara Arnaud of NASA, a specially designed space trampoline might. In her special study of the part hormones play in osteoporosis, she suggested that "there's some genetic connection too. For instance, blacks tend not to get osteoporosis, whereas, post menopausal, blue-eyed, blonde Englishwomen are the classic victims."

Of course, the permanent space resident who never plans to set foot on Earth again would not have to sweat and exercise an eighth of his life away, or gobble down vitamin supplements, or sit under hot lamps. "The aim of space medicine would certainly change to defining and maintaining what good health would be then," said Oleg Gazenko, head of Russian space medicine. "The body would seek and find its own state of homeostasis. The object of space medicine will be to keep the body healthy in that environment."

If space stations are not built to rotate and provide pseudogravity—and there are compelling reasons not to—the legs, heart, and certain muscles would atrophy since they would no longer have to fight weight. Someone living on the Moon would physically resemble a space-station native since the Moon's gravity is a negligible one-sixth of Earth's. However, Martians would be slightly more fit, since that planet has 40 percent of Earth's gravity force at its surface, enough to compel certain muscles, the heart, and the bones to work harder.

Some space medical experts have pointed to other, possibly far more profound, changes. When large space populations are exposed to radiation, it's possible that the biological basis of our species will be changed through genetic mutation. That may or may not be true. The Kerala province of India sits over a big uranium field. People there are constantly exposed to low levels of radiation. Although they have a somewhat higher leukemia rate, there is no noticeable increase in mutation. Nor did the survivors of Hiroshima and Nagasaki give birth to any such legion of genetic changelings.

Monstrous genetic mistakes—human creatures born without a brain or with hideous masses of flesh or with one cyclopic eye in the middle of the forehead—are usually spontaneously aborted or die soon after birth. A large number of birth defects arise not from some genetic problem but from the mother's exposure to chemicals or illness during crucial times in fetal development. Toxic chemicals will be a fact of life in space, as on Earth. An intelligent, cautious, responsible mother can usually prevent such defects, except in the case of unavoidable accidental exposure.

Natural selection will play a part in space, and may over many generations kill off certain inheritable traits. What characteristics lead to survival in space? Certainly, tolerance to higher carbon dioxide levels, the ability to flourish on limited water and food, and insensitivity to radiation would all be useful traits. It's difficult to say if radiation sensitivity is inheritable, although there's some suggestion that certain metabolic characteristics can run in families. Even if radiation sensitivity were inheritable, it might take many generations to cull out that trait, if ever. Many life-threatening traits can still be perpetuated provided the traits don't kill in childhood; one would expect hemophilia to be vanquished by now. The same might be true of radiation-induced cancers in prone individuals: it won't kill them until they've had a child or two.

Natural selection is a very slow, albeit directional process. It took millions of years for us to develop as a species, and we haven't changed noticeably in tens of thousands of years. It's not likely to be any more rapid in space.

Mutations are faster, but are nondirectional and unpredictable. We tend to think of mutations as terrible and monstrous because they are out of the natural order of things. But evolutionists like Harvard's Stephen Jay Gould point out that mutations allow for leaps up the evolutionary ladder. Without them, necessary evolutionary changes might occur too slowly for our survival. Species would stagnate.

People don't think of mutations, natural selection, the dangers of life, the certainty of decrepitude, or the horror of death when they have children. It is only fitting to end a chapter about death in space with the most important feature of the future space settlements: childbirth.

The mother would have to be strapped down to some surgical table, and the doctors and nurses anchored in place around her in preparation for the delivery. She would be prepped in the time-honored way. A plastic bag might be inflated around the lower half of her body to create a sterile field and to keep blood and fluid from spreading all over the room. The bag would contain all the surgical tools the doctor would need, built-in gloves and arm portholes for nurses and doctor, and inserts for suction and water flow. Childbirth is a messy process, and nurses would have to be assiduous in using both suction and water to keep the sterile field and the plastic bag free of debris.

Contractions proceed of their own accord and are not gravity dependent; they will not be easier or harder because of the space environment. Labor might be slightly longer if the space babies do not drop down toward the birth canal as they do in the last weeks of pregnancy on Earth. Delivery is also not weight dependent, since there have been numerous underwater deliveries with no harmful effects.

The baby would emerge from its mother's womb into a plastic bubble. Its mouth and nose would be suctioned, as usual. Its first cries would not be as sharp and piercing as in an Earth delivery room until it was removed from the plastic bubble. The doctor or nurse would carefully move the child to a section of the bubble that could be closed off, in order not to break the sterile field around the mother. As on Earth, the doctor would work on delivering the placenta and sewing up the episiotomy or natural tear. The bubble cannot be taken down and the sterile field cannot be contaminated until the surgery is complete.

The mother asks the universal question: Is the baby all right? Yes. There has been nothing so far in space research to suggest that the baby would be other than all right.

The baby would be handed or even floated in the direction of a nearby receiving area where it could be cleaned up. The move from the weightless environment of the womb to the weightlessness of a spacecraft would probably be comfortable and familiar, almost like a LeBoyer birth without the water.

As on Earth, the ritual of certification begins. The bottom of the feet are inked, and the baby's footprints are pressed onto a birth certificate that must be clamped down in order not to float. The mother's fingerprint is placed beside the footprints. The baby bleats pitifully until the certification and the bath are over, while the doctor, the nurses, the mother, and possibly the father chat in a casual fashion.

The baby is swaddled and handed to the parents. It is hard to keep hold of the baby since it floats at the slightest loosening of the parents' grasp. In this environment the mother feels as if she is holding nothing but air. The swaddling blanket takes on a life of its own and comes undone. It is secured with Velcro straps to keep it from unraveling completely.

The mother can look forward to some troublesome aspects of raising a space child. Diapering will be a bit of an ordeal since it is hard to hold the baby in one position (weight takes care of this automatically on Earth). Changing tables will have to come equipped with padded baby vise grips or straps to keep the infant secure while being changed. If the baby squirms, which most do, even a tiny thrust of its leg is enough to propel it away. The American Indians had the right idea when they invented a papoose sack in which the child can sleep and be carried around; it's the only practical way to deal with a baby in space since it restricts the child's movements and conveniently secures it to the mother. A little helmet will be a must: space babies will undoubtedly get loose from time to time and smash their heads against a wall. Babies will be much more mobile much earlier in space since they needn't learn to crawl or walk to get around. That frantic stage of childhood which all mothers dread in which the child is mobile but not sensible enough to keep itself from disaster will start earlier in space and last longer.

The child's citizenship would probably be that of the mother. The place of birth is somewhat more difficult. It could be "longitude 95 west, latitude 30 north, altitude 312 miles," or some established point in space: Lagrange 4. It could be the name of the space station: Vonbraunberg, Goddardton, Korolevsk. Or, because the spacecraft is not a fixed place on a planet, it might simply be "Space."

The name on the certificate must be filled in. Maybe it will be something grand to commemorate its space origins—Luna, Stella, Venus, Cassiopeia if it's a girl; Orion, Arcturus, Cosmo, or Jupiter if it's a boy. The parents might think it only fitting to name children after the planets and stars that are their home. Or they might name the child after some special place on Earth: Francesca or Francis, Geneve, or just Paris or Houston. The child might be named for an idea for which space stands: Chastity, Prudence, Liberty, Irene (which means peace), or Zoë (which means life). Or parents could carry Earth names with them wherever they go in the universe: Gregory, John, James, Alexandra, Sally, Mary or, for that matter, Sven, Vladimir, José, Niko, Guido, Kurt, Britt, Tatiana, or Gabrielle.

The most appropriate name would be that which a human birth in the black, death-dealing void of space truly represents for mankind. Hope.

chapter ten

MENTAL LANDSCAPES

I once thought of writing a novel about a long voyage in space in which one of the characters on the ship was a son of a bitch. He was a villain, and I was going to have him all through the book. He manages to avoid being crushed by various hateful dodges and, at the end he gets off the ship, and they give him a medal because this is the first long voyage that people had gone through without going crazy through all the loneliness of it, and the weariness of it, because they had something to hate, something to fight.—*Isaac Asimov, 1977*

Just because you're paranoid doesn't mean they're not out to get you.—*poster, late 1970s*

Maturity is the ability to get through life without making other people sick.—*Erich Lindemann*, Beyond Grief

The opening of a frontier is nothing less than a test of human sanity. Sea voyages across uncharted oceans and wagon trains across unfamiliar prairie suffered psychological casualties along the way: mutinies, violent brawls, murders, deep depressions, and even psychotic episodes. By comparison, the initial stages of our outward movement into space have been incredibly and mercifully easy.

Part of this is undoubtedly due to the rigid control of the types of individuals chosen to go into space. The Mercury astronauts submitted to twenty-three psychological tests prior to spaceflight selection. At that time, the space program wanted intelligent (especially in mathematical and spatial functions), reliable, consistent, deliberate, self-confident, and motivated men who were calm in an emergency. Having obtained a cadre of exactly that type of man, NASA must have been reminded of the old proverb: "Be careful what you wish for— you might get it."

In later years, when the manned space program's aims changed, so too did the image of the ideal astronaut. Intelligence, motivation, and level-headedness in an emergency were still prime requisites. But now, the astronaut personnel office preferred cooperative, socially compatible, well-rounded human beings, with the emphasis on cooperation.

Psychological tests had been dropped from screening procedures in 1977 because they were deemed ineffective gauges of human personality, judging from the later history of some of the astronauts who had "passed." However, current American would-be spacefarers still have to go through a test for claustrophobia, a psychiatric interview, and some stomach-wrenching airplane aerobatics before they can be picked to ride aboard the shuttle.

Dr. Kirmach Natani, an expert in selection procedures, has claimed that selection for isolation duty based on psychiatric and psychological interviews is not as successful as hand-picked crews assembled by veteran leaders of isolation duty. Neither Russian nor American spaceflight commanders are allowed to name their crews, although a commander can put in a good word for someone. Crews are named by flight-operations directors, based on seniority, specialization, and intuition.

But in selecting people for long-duration flight, the Russians are much more wary, and put their cosmonauts through more stringent and sophisticated psychological screening. All American space-shuttle flights (and most Soviet orbital visits, too) last only about a week, and one can put up with just about anybody for that amount of time. But some Soviet orbital expeditions last half a year or more. As he rode the bus to the launch pad, Valeriy Ryumin mused: "O'Henry, the

American writer, wrote in one of his stories that if you want to encourage the craft of murder, all you have to do is lock up two men for two months in an 18 × 35 foot room." The interior of the Salyut is only about six by eight by thirty feet. The consequences of incompatibility and conflict among crew-members could be extreme.

Russian spaceflight psychologists A. Nikolayev and M. Gor-yachev have described the Soviet process for unmasking in-compatibility. In a case study, they found latent hostility even in "two comrades [called Valdimir and Nikolay] who for eight years had lived and worked in a single team of test specialists.

"Vladimir and Nikolay were invited to sit at instruments from which a wire extended to a machine in the corner. . . . The doctor started the machine. The instrument hands were immediately deflected from the zero position. Vladimir and Nikolay, each in his work place, began to turn levers in order to turn the hands to a zero reading. Vladimir turns the lever of his instrument, the hand moves on Nikolay's instrument as well, but his toward the center and his comrade's toward the edge of the scale. For a long time the two friends 'pursued' each other with the hands, but they could not set the two instruments simultaneously on zero, as was required by the physicians. Vladimir began to be nervous, cast angry looks at Nikolay, and the latter, frowning stubbornly in his own fashion, turned the levers. The hands first approached the zero readings and then departed from them."

This is a description of the "homeostat" test, which has been written up extensively in Soviet psychological litera-ture. The two cosmonauts must coordinate their movements to bring an indicator needle in an instrument board down to zero. Their attempts are frustrated if they try to disregard each other. Even for those who succeed at some degree of coor-dination, the effort becomes more difficult as they go on. According to Mikhail A. Novikov, a laboratory head at the Institute of Biomedical Problems and an important figure in Soviet space psychology, "In a cooperative activity [the hom-eostat], there are frequent situations where the subjects counteract each other in the 'struggle for the ultimate way' with pronounced behavioral reactions—fidgeting, remarks against the partners, disapproving looks in each other's direction, and so on. The activity loses its cooperative nature and has a distinctly competitive character."

In another test, called "pair matching," word stimulators are rapidly generated and the two subjects must compete to beat each another to the right answer. During this forced competition, the unconscious physiological responses of the nervous system are measured, to give psychologists an independent indicator of emotional state. Called "neurometric assessment" in the U.S., the medical monitors include pulse, respiration, brain-wave activity, and galvanic skin response. Again, Novikov noted that successful pairs of cosmonauts overcame the problems of forced competitions: "The subjects frequently depart from competition, join forces to struggle against circumstances and arising difficulties."

The Russians carefully noted that reaction time varied greatly among tested pairs. How a cosmonaut tuned out interference from outside and from his partner, and what his body posture expressed were also of interest. Factors that seemed to influence compatibility included intellectual ability, skill, a positive attitude toward the activity and each other, mutual empathy, learning ability, adequate distribution of work and responsibility, and uniformity of physical responses.

Next the cosmonauts are locked in airtight chambers—one representing the spacecraft, the other the space station—for periods of ten days to three weeks, simulating a real spaceflight.

On the first day, test subjects Vladimir and Nikolay are made to sit in their spacesuits, simulating launch problems until their knees go numb. After three days of intensive work, their pulse rates drop, they become fatigued and stressed. By the twelfth day, noise levels, physical discomfort, and fatigue have piled up. The cosmonauts snap:

"Leave the trainer alone!"

"What's with you? Don't you have anything to do? Busy yourself with something else and don't bother me."

"No, you do something else. Who will change the filter in the analyser? Me, huh?"

"And why not you, comrade big boss?"

Nikolay replaces the filters, and sleeps in the transport ship, not in the station as he did previously. These are two men who are naturally competitive, a terrible risk for a long-term space voyage.

After extensive research along these lines, the Russians found that crews tended to break down into one of three characteristic types: "congruent," "complementary," and "competitive."

A congruent spaceflight team consists of individuals who are similar in personality and can function interchangeably at the various tasks. There is a quick acceptance of roles in this group, in which the leader-follower relationship is quickly established and defined.

In complementary teams the personalities of the teammates are different. "Contrary to hitherto accepted notion," said Oleg Gazenko, head of the Soviet space-medicine program, "people with 100 percent different personality traits proved to be compatible. An open-hearted, easy-going and sociable person might be at ease with a reserved, taciturn and private person. It is important that they coordinate with each other emotionally and intellectually."

The Russians discovered that cosmonaut compatibility even went as far as the heartbeat. "In the course of the paired verbal test," Gazenko said, "the pulse of partners helping each other synchronizes. When the test is interrupted, pulses desynchronize." The question of body function synchronization has been raised in the West, where studies on college students' menstrual cycles suggest synchronization among roommates and friends. Whether such things as heart rates and menstrual cycles can be indicators of psychological compatibility has yet to be proven. The Russian space psychologists seem to swear by them.

During spaceflight, the Russians found that the closer the pulse rate, the better the cooperation between the cosmonauts. Cosmonaut teams consisting of Lyakhov with Ryumin, and Kovalyonok with Ivanchenkov had very similar pulse rates in preflight tests, and on their long-duration expeditions they seemed to do well.

Novikov and another Russian researcher, V. Lenskiy, found that congruent groups, consisting of people of similar personalities, performed inconsistently in problem-solving situations. "The presence of rigid structures in congruent groups who strictly follow the leader-follower concept negatively affects problem solving," Novikov reported. "Group congruence is desirable in conducting short-duration tasks, but is characterized by a certain behavioral rigidity, a lack of flexibility in solving tactical problems."

People who are too similar can get on one another's nerves too, apart from leadership structure and problem solving. There is some suggestion that Valentin Lebedev and Anatoliy Berezovoy, on their 1982 seven-month mission, were too similar. They were both persistent, thorough, intense nitpickers. In his published memoirs and in subsequent professional papers, Berezovoy proposed sending teams with dissimilar, complementary personalities. "It is advisable to form crews in such a way that the merits of each member are complemented while the defects are smoothed over," he wrote. "If this is done, a crew will be strong and capable of going further than we did."

Aleksandr Ivanchenkov, a veteran of a 140-day mission in 1978, provided corroborative testimony about this effect. "Outer space is filled with surprises," he told an interviewer in 1984. "These surprises can be dealt with only by acting in concert, by silently understanding one another. I might mention that Kovalyonok and I are completely different types. He is the kind who plans things out, who reacts to things slowly. I am more reactive and more sure of my actions as a result of having been involved in downhill skiing for more than twenty years. You'd be surprised, but this difference of characters only helped us get along."

The Russians found that although such complementary teams

tended to need longer training, they ultimately proved their worth. Novikov called them "the most reliable for protracted group functioning in extreme conditions. . . . [This] stimulates creative activity in partners and contributes to arriving at optimal solutions in confused situations." Also, the arrangement provided "support group unity in cases of monotony." In essence, such teams were more flexible in their problem-solving tactics, and tended to make up for one another's deficiencies.

In late 1983, *Izvestiya* correspondent A. Ivakhnov claimed that the ideal combination was achieved by Vladimir Lyakhov and Aleksandr Aleksandrov. He profiled them this way: "Vladimir Lyakhov is a well-known optimist and a good fellow. In any situation he quickly grasps the essence of it, stripping away the trifles and isolating the main point. He is not afraid of risks and makes decisions bravely and decisively. Aleksandr Aleksandrov is just as industrious but his nature is entirely different. He loves to dig into petty details and in every question he tries to dig down 'to the very bottom.' He does not hurry when making decisions, but weighs all the factors and conditions and compares every 'for' and 'against.'"

Lyakhov had also flown an earlier successful long-duration mission with his opposite, the reserved worrier Valeriy Ryumin. Ryumin, like Aleksandrov, approached situations with caution, keenly aware of inherent dangers and problems.

The Russian cosmonauts are somewhat less standoffish about sharing their feelings than are American astronauts. After a scientific exchange visit, American space doctor Joseph Sharp said that "The American-doctor-to-Russian-cosmonaut talk was better than an American-doctor-to-American-astronaut talk." Another psychologist described his impressions this way: "The astronauts seem to act like I'd expect Russians to act, and the cosmonauts seem to act like I'd expect the Americans to act. That is, I think we should beat them on such dimensions as openness, honesty, candidness, and general interpersonal sensitivity—and we don't. Instead, we seem to be the unemotional taskmasters (or taskmistresses), and they seem to be the touchy-feelies."

Russian spacefarers repeatedly counsel each other informally on psychological matters, an impossible dream for American spacefarers with their traditional hostility to all things psychological. For example, Berezovoy recalled that Popov and Ryumin had advised him that "constant occupation and work, patience and an honest desire to understand" helped smooth over the crises in relationships that always arise in space. During a short visit aboard the Salyut in 1982, Aleksandr Ivanchenkov, a long-duration veteran, casually mentioned to Berezovoy and Lebedev that based on his own experience, depression was absolutely normal during long spaceflights.

Prior to flight, some cosmonauts made pacts to work together to solve interpersonal problems. Berezovoy wrote: "During training, we set outselves a target: we would map out all the 'hidden rocks' that we would perhaps bump into in orbit. . . . I am no longer a lad, and everyone has his convictions and habits, his own style of work. And sometimes they do not coincide. . . . We immediately agreed that in the work there, there should be complete frankness." Said his crewmate Valentin Lebedev: "We swore an oath: in any situation we would be temperate, good-willed, try to maintain good relations."

According to the spaceflight diaries of Ryumin, Berezovoy, Lebedev, and Aleksandrov, the greatest psychological challenges seem to lie, somewhat ironically, mostly in the areas we associate more with crowded urban living than pioneer questing.

Ryumin pointed out that training took place in the presence of others, and one really did not get to know one's crewmate until they were alone together. Lebedev and Berezovoy felt as if they knew each other very well prior to their flight—that is, until they flew. Said Berezovoy: "Valentin and I had a very short and very intensive period of training together. And so some questions of our mutual relationship were simply not solved completely successfully. We had to deal with them in space."

In spite of all this psychological screening, Russian space crews had their problems, although nobody has been specific about precisely what the problems were. There are limits to Soviet space candor.

Although some in Russia swear by the psychological compatibility testing and the heart synchronization measurements, not all do. Quite often, external conditions or problems dictate that crews be changed around, as happened when Ryumin substituted for Lebedev only a few weeks before a long-duration spaceflight when the latter injured his knee. In 1977, a replacement of one flight engineer by another with a badly needed special repair skill caused a crew assignments ripple down the next three flights (each flight engineer got bumped one crew downward), with no detectable adverse psychological effect.

Cosmonaut Vladimir Dzhanibekov pointed out that compatibility might just be the peculiarity of a certain crew. "Now that I have returned to Earth," he said after a short flight in mid-1984 during which he observed a crew that had been aboard Salyut for five months, "I want to tell the psychologists that there may be no problem of compatibility at all."

Gazenko too warned that psychological compatibility was only a part of crew dynamics: "Good relations are dependent not only on psychological compatibility, but also on the individual qualities of man. Remarkable results come from patience and restraint. Unplanned situations can always arise: there can be delays in the receipt of necessary information; there can even be errors in actions of the partner, but it is always necessary to be able to restrain oneself, not to show irritation which can be conveyed to your partner and only worsen matters. Any heating-up of the situation will also decrease your performance and your feeling of well-being. Evenness of behavior, capacity of tolerance of physical and psychic loads in marathon fashion for a long time—such are the qualities necessary for a prolonged spaceflight. These qualities, like the necessary feeling of comradeship, are instilled during the course of man's entire life and constitute an integral part of his culture and breeding. In short, the culture of individual behavior and communication with one's flight partner and the operators at the Control Center, with an increase in the duration of space flights, is acquiring an ever-increasing importance."

Groups of three people worried the Russians at first. Triads tend to break down into a pair and a social isolate. In the U.S., a long-running experiment at Johns Hopkins tested how triads behave under isolated and stressed circumstances. The groups made mistakes, made negative remarks to outside researchers, broke equipment, went on strike, and one person even began to hear voices (although this individual had had previous psychiatric problems he had failed to mention to the researchers during screening).

They sometimes scapegoated the isolate. If the social isolate was a loner type, and really wasn't bothered by being ostracized, the psychological fallout was negligible. But most people are affected by hostility and being cut off from normal human bonding. What worried psychologists about sending triads into space was that in these extremely sensory-deprived, isolated places, even tightly knit groups are exposed to severe psychological strain. The isolate, with no other social options, can become withdrawn, depressed, and may even experience a mild form of psychosis called "long eyes syndrome."

Connors explained: "This is what you would call a twenty foot stare in a ten foot room." The ostracized person rocks back and forth and just stares into space. However, this is a temporary psychosis that disappears as soon as the group feels sorry for the distressed individual and accepts him or her back into the group. In other primate societies, a very similar gesture is regarded as a form of submission, and the dynamic of reacceptance is similar.

In spaceflight, however, the triad grouping turned out to be quite different from what theorists had predicted based on earthside analogies. The reasons for this are manyfold. For one thing, anyone who goes into space is carefully screened for mental stability and high motivation, which go far to overcome many psychological problems. Also, in space, the strict tendency is to suppress hostilities. American Skylab crews always consisted of three men, and everything seemed to work out well. The groups did not break down into a pair and social isolate in the classic sense.

Groups of three offer important advantages for spaceflight. If one member is feeling ill, there are two others to pick up the slack. Obviously there are more social options in a group of three than of two: one can socialize or withdraw more easily and with a different result here than in a pair.

The 1984 eight-month Soviet orbital mission of Kizim, Solovyov, and Atkov seemed to prove this. The bright but relaxed Kizim was teamed with the high-strung Solovyav and the physician Atkov, who had also received special psychological training. For the early part of the flight, Kizim played the role of a hard charger, and Solovyov was the retiring, meek documentation keeper. Later, Kizim (who at first strikes people as lazy and flabby, but then surprises them with sharp intelligence) developed several major fuel-saving maneuvers for supporting scientific activities. Vladimir Dzhanibekov, who visited them for twelve days and witnessed their interaction firsthand, thought the psychological climate "amazing"; he later told newsmen, "I would describe it as ideal for the long, interplanetary flights that will certainly take place at some time.... Their mutual supervision, help and mutual support ... may not be very explicit but can be traced very well all the same.... This atmosphere, this favorable human climate, helped us considerably in completing the entire program."

American space experience with three-man crews was also remarkably positive. Another doctor, Joe Kerwin, was a crewmember aboard Skylab for twenty-eight days in 1973. Kerwin (who writes detective fiction as a hobby) described the psychological climate this way: "It takes awhile for a crew to fall into a routine, neatly dove-tailing with each other. When you rush 'downstairs,' late for a medical experiment, and find that your partner has already done your part of the prep, you know it's working. You know you've got a smooth and very versatile machine, almost infinitely reprogrammable, not error free, but self-correcting, learning from its mistakes. A machine that gets the giggles late at night over ice cream and strawberries, and occasionally looks out the window when it could be writing the log, but which lives and is replenished by these things, better than a machine is by its voltages and lubricants, and will be ready to go again tomorrow. Real teamwork is memorable, and in space, it's just the same. People perform up there the way they do down here. Their capabilities, individually and collectively, and their potentials, and their weaknesses are the same. And when we're ready to dispense with people on Earth, we can do it in orbit and the computers can talk to each other forever."

Psychologists intent on defining the psychological profile of future spacefarers have pointed in the direction of people with extraordinary psychological flexibility: the so-called "ambivert" and "androgyne." The ambivert is defined by Connors, Harrison, and Akins as "a person who exhibited behaviors commonly associated with both extroversion and introversion." In theory, this person could overcome the contradictory social demands placed on a person in space—to interact with one's crewmate without driving him or her crazy with constant pressure to socialize or break the ice of silence.

The psychologically (not sexually) androgynous individual would also be a desirable blend of traditional opposites: intense goal orientation usually associated with the male, and interpersonal sensitivity commonly associated with the female. University of Texas psychologist Robert Helmreich argues that space-station living requires high degrees of professionalism and goal motivation to accomplish the difficult and sometimes boring tasks, while a high degree of social sensitivity is required to ease problematical crew interactions that always attend such close confinement. The psychologically androgynous people may be the answer to the call of space.

There is, in fact, some evidence that this type of person is already emerging in both the Soviet and American programs. For example, Leonid Popov, who flew with Ryumin on his second long-duration flight, might have that professional-social blend which makes him a kind of "spaceman for all seasons." Ryumin wrote: "I liked Lenya [Popov]. He related well with the rest of the boys in the cosmonaut corps. . . . [He is] calm, judicious, with a certain gentle warmth. He showed a desire to resolve debatable issues with the least amount of loss to anyone. He was respectful among his friends. Later even in training and in the flight, I was convinced that Lenya was highly skilled in resolving technical problems, adapting quickly and correctly in the subtleties of our uncommon technology." In flight, Ryumin praised their "wonderful relationship without offense or reservations and with constant joking and good-humored drawn-out playfulness," along with Popov's professionalism: "I liked the fact that Lenya, flying for the first time, did not feel himself a rookie and from the very first day dove into the work. As far as I was concerned, this kind of behavior merited the highest praise. On Earth, they were pleased with the great volume of work which had been accomplished." This is high praise from the normally reserved Ryumin, but others must have reached the same conclusions about Popov's space leadership style. Although billed as "the kid" (he was the first Russian cosmonaut born after World War II), he was quickly given two more flight assignments to become the youngest man ever to have flown in space three times, having commanded three Soyuz spacecraft before his thirty-seventh birthday.

During an interview after their flight, Lyakhov and Aleksandrov were asked what are the most important cosmonaut qualities. "Kindness and professionalism," they answered without hesitation. The response points to that blend of the human and the work-related dimensions of space which can be termed "psychological androgyny."

Driven achievers have, for the most part, been the rule for astronauts and cosmonauts. And it would be a mistake to expect "laid back and mellow" to be the wave of the future, especially in the relentlessly hostile space environment. "Space is a strict taskmaster," wrote Berezovoy. "I remember that some comrades started to take offense at me for allegedly great faultfinding, but I have already realized that the attitude of forgiving your partner for everything only hampers things. I was constantly worried by the thought that many of the operations in the program were having to be done for the first time. If any deviation occurred, it was essential to analyze the situation rapidly and make the correct decision."

In all likelihood, the spacefarers of the next hundred years will all have to be calm and level-headed in an emergency, and competent at their work. "It wouldn't hurt to choose someone who is somewhat paranoid," claimed Kirmach Natani, "someone who continually takes into account the potential hazards of one's environment. We don't want risk-takers who see danger as a kind of cleansing by fire!" "Maturity" is an ill-defined quality that is considered ideal by psychologists and astronauts alike. "It implies a mellowed emotional response, experience, and wisdom," said Connors. A knack for survival would not hurt either.

Spaceflight psychologist Albert Harrison suggested that another factor needed consideration: he termed it "the annoyance scale." It could be used in two ways: "One, to eliminate people with too many annoying traits and mannerisms, and, two, to eliminate people who found too many traits and mannerisms annoying!"

But beyond that, it is a vain endeavor to draw up a profile of a typical future spacefarer, since many different professional and social niches will appear in space. Each will require a different type of person to fill it; there will be no ideal type to fill them all.

At a glance, a space station microsociety might have niches for as many as six very different types of people. The most important is the leader. As it stands now, power aboard a spacecraft is vested with the crew commander who, in turn, takes his orders from the flight director in Mission Control in Houston or Moscow. The buck stops there for all practical purposes. However, with the advent of "payload specialists" and "cosmonaut-researchers"—nonprofessionals (and even nonnationals) who fly on the shuttle and on Salyut for scientific, technological, or other reasons—the problem of leadership and authority becomes much more complicated.

On a Spacelab mission, for example, scientific experiments are overseen by a NASA astronaut called a mission specialist. The payload specialists hypothetically answer to this mission specialist. But ground-based scientists who designed the experiments also have something to say about how the experiment is conducted.

With military space shuttle cargo, the payload specialist must answer to the mission specialist, the commander, and an officer on the ground. One can already see the ingredients of an authority nightmare in all this.

For space station operations, most people agree that there should be a blend of "town-meeting" democracy and benevolent despotism. A commander can make the ultimate decisions on the spot. Although for a long time to come there will be flight directors who will direct the space mission and coordinate the support services on the ground, the space station commander will have to have more autonomy since the remote exercise of authority is always a problem (the "wait till your father gets home" syndrome). In an emergency, there isn't time to vote; in threatening circumstances, instant compliance under a unified leadership works best.

NASA laid the groundwork for the future by establishing a rule whereby the spacecraft commander has the legal authority to use "any reasonable and necessary means, including physical force" to maintain order. Hypothetically, he or she can make an arrest and charge someone with a crime. However, big crimes (those with large fines, large prison sentences, or capital crimes) would probably have to be adjudicated on Earth. In essence, the commander has the legal authority to punish a brawl in space without recourse to authorities on Earth, but he or she cannot hand out a death sentence.

In everyday circumstances, democracy appears to work best. On Soviet spacecraft, there is always a commander, but he still has to clean up the station and cook some meals along with everybody else. This follows an old rule in isolated and hostile circumstances: All are equal in the struggle against nature. More seriously, small-group cohesion is threatened whenever an individual is exempted from unpleasant tasks merely by virtue of a title.

So far, Russian long-term missions have been run on democratic principles, although several crew commanders (who have always been military jet pilots) complained that mission controllers tend to ask questions of the flight engineers (who have always been civilian engineers, often former flight controllers) rather than the commanders. The commanders are, after all, responsible for all the work aboard the station. Nonetheless, some crew commanders, notably Popov (who was a rookie when he flew with his older and immensely more experienced flight engineer, Ryumin) have ignored the question of command altogether, thus eradicating a potential point of friction in a very long expedition.

Qualities of leadership vary with the needs of the group. Psychologists have found that severity in a leader is tolerated, but not incompetence. Harrison described polar expeditions where weak, indecisive, and incompetent leadership spelled disaster. In the Antarctic and other isolated environments, long experience was also favorable.

Good leaders, like good managers, have the charismatic

ability to call up the best from the crew, even to the point of self-sacrifice. Connors, Harrison, and Akins point out that in space, where traditional bribes and threats are ineffective, personally severe leaders may have the advantage in the short term, and pleasant, socially skilled ones in the long term.

Both the Soviet and American space programs have had to deal with the "emergent leader" who can sometimes challenge the authority of the "assigned leader." The assigned leader may direct the mission tasks well but be unequal to those interpersonal pressures with which a sensitive, intuitive crewmate—a second or "emergent" leader—might cope. This emergent leader is never assigned that position—he or she just assumes it as a natural exercise of his or her innate skills. The usual explanation for the emergence of a second leader is that the task leader's sense of purpose gives rise to activities (unpopular orders, sharp criticism, etc.) that hurt group members' feelings. The second leader emerges to smooth things over and restore harmony to the group. Enter Mr. Roberts, or Mr. Christian.

The "boundary line" person has been the subject of interest in isolation/confinement studies. The "chief of the boat" or chief warrant officer aboard a submarine is such a person. "This person is both a counselor and a disciplinarian in small matters," said Jack Stuster of Anacapa Sciences, whose NASA study surveyed the best analogs to long-term spaceflight. "This person prepares the day's schedule according to the disposition of the crewmembers, and acts as the ombudsman to the leadership."

The boundary line person may be assigned this task of go-between, or may just take on the role. The communications officer may be such a person on a space station: "This crucial person would serve as an interface between the outside world and the inside group," said Mary Connors. "Because this person is placed literally between two worlds, the burdens are very high, and he or she receives pressure from both sides. This person is usually very cautious and tactful, and is not trained but rather intuits his/her way through situations. But the function is crucial, a key interface, and one that cannot be ignored or remain unacknowledged in the future as it is now."

The strain on current space crews has been precisely at this juncture—the Earth-space connection. There is the ever-present tendency of people in a confined situation to direct hostility outward. This has already happened in both Soviet and American programs. In one case, a Russian communications officer got into trouble for cutting across the lines of command and by directing the communication to the scientist-cosmonaut aboard the Salyut rather than to the commander.

Also, people on the ground do not always appreciate or understand precisely what is going on in space, giving rise to the potential for misinterpretations and mistakes. Such an episode happened aboard Spacelab 1 where ground scientists kept requesting changes and additional tasks of the space crew without realizing that the crew was already deeply involved in ongoing experiments and couldn't easily do what was requested without disrupting that work. Astronaut Bob Parker finally had to lecture the ground severely about constantly being asked to drop one job and turn to another. This fragile line of communication with Earth is fraught with such tension and problems; it will become an even larger burden in the future.

Probably the most neglected role in society, and the one the cosmonauts already recognize as the most sorely missed in space, is that of the nurturer. This person's unacknowledged contribution to society helps make human life human, makes home home. Neglected in space psychological literature as in society as a whole, this person has the ability to make a space mission a pleasant kind of living rather than an extended camping trip. In the Antarctic the nurturer is usually the cook who, along with the communications officer and the commander, has the highest status. Food has such fundamental associations with home and family that it would be foolish to rely indefinitely on ready-to-eat, preprepared meals when one can have a morale-boosting cook aboard to surprise, delight, and nourish all at once (even a part-time cook might do).

A nurturer need not be a cook, of course. This role could be filled by a particularly sensitive doctor or a loving, caring person who enjoys giving of himself or herself to other people. Space will be an empty place without those fundamental mother and father figures who are scattered throughout our lives on Earth.

The scapegoat in space will undoubtedly be the person whom the group perceives as incompetent, a bungler, a maker of mistakes, or just the person who is slightly different from the rest, possibly one who consciously or unconsciously challenges their values. The pressure to conform in a small society in a hostile and isolated environment is intense. The scapegoat might receive the collective blame for real as well as perceived problems. There are people who may not be bothered by being the scapegoat since they have always been social isolates to begin with: the "wimp," the "egghead," the "nerd," the "jerk." Fortunately, the annals of art and science overflow with such individuals, who sometimes, in the end, provide a culture with its greatest boon: insight, genius, art.

In fact, such people may accept the label and the role of scapegoat willingly since it conforms exactly with their own self-image, the part in which he or she has already cast himself or herself. The danger in this is that in a space habitat there is no social or physical exit. If there is no psychological buddy—some other nerd or wimp—with whom one can socialize, and if the scapegoat does not have the immense human fortitude to take the barrage of ill feeling day in and day out, then a nervous breakdown, a psychotic episode, or a murderous rage are distinct possibilities.

The clown, such an important figure in any group, has also been virtually overlooked in space psychological literature.

The clown might be defined as someone who is never quite taken seriously, but who has an unacknowledged role in keeping morale up. The clown can dispel tension, lessen attention to things that become disproportionately important, and generally help the group keep its psychological balance. In a way, the clown can challenge "groupthink," the tendency in all small groups to conform rigidly and get into a mental rut. Because he or she is the jokester, this individual can challenge groupthink with impunity and possibly without becoming the scapegoat if the clowning doesn't carry over into professional incompetence. In the Middle Ages, kings had fools and court jesters who entertained and sometimes criticized. Naturally, in space the clown will emerge from within the group and have to intuit just what the target of his or her lampoons and jests might be. Eventually, space communities will be large enough to hire professional comedians to satirize their society, to cheerfully hector the leaders, and to poke fun at what makes them successfully and crazily unique.

Space already suffers from too much sameness. Astronauts are supposed to be fearless, intelligent, competent, motivated, and dedicated. But they are also, so far, a pitifully inarticulate lot who generally see life with an engineering, flying, or scientific squint. That's fine for now. Similarities in views, attitudes, backgrounds, education, and training are powerful binders of social cohesiveness, and are the building blocks of successful microsocieties. According to Connors, Harrison, and Akins, they have a relationship that goes from the past to the future and are willing to sacrifice short-term gains for long-term ends. But this similarity cuts down somewhat on the diversity and enrichment of life, and possibly on new approaches to both technological and social challenges.

Already people outside the tight elite corps of astronauts are being allowed to fly in space, although only a few and only one by one. These artists, writers, teachers, politicians, philosophers, and so on will enhance the space experience by being there, by having taken the less-trodden path that leads to insight, beauty, or wisdom. When, in twenty years or so, we cease to speak about "space missions" or "space flights" and talk of "space societies," these people will be even more crucial in giving shape to the amorphous and the alien, sense to the senseless, direction to the random and multiple experiences of the future. "We cannot leave space in the hands of the technologists," said James Michener. "It would be a terrible loss if we did."

In the early extraterrestrial microsocieties of the 1990s, when space-station populations are small (only a dozen individuals at the most), alliances may form along marital or professional lines or personality similarities. Scientists may form a clique apart from the NASA command crew or the military contingent (if any), as they sometimes do in the Antarctic. The oddballs or nerds might run in a small pack. Women might form a support group within a group, as they did on most of the wagon trains traveling across the American West in the late 1880s. Musically inclined people could relax by playing in a band. Married couples could share their problems with one another, since some of their pressures are different from those experienced by single people.

Such groups may be a source of conflict and divisions of loyalty, but they also work to enhance each individual's life in space, diversifying the number and kinds of roles one can have. And, as Connors, Harrison, and Akins pointed out, controlled conflict is useful. Conflicts alleviate depression, synthesize the best solution to a problem, and draw attention to problems that might otherwise be left alone to fester. The last Skylab astronauts had to air their grievances to mission controllers during their flight or their insupportable work schedule would have continued, and the intolerable pressure on the crew would have gone unalleviated.

So long as everyone has a place, and cliques don't become islands unto themselves that threaten the unity of the mission, it's okay. That's the verdict of recent experience in orbit, and of analysis on Earth.

Ultimately, though, a group's success will not even be a question of excellent prescreening, good professional training, adequate role-playing, or a combination of those. Groups are made up of individuals. Some groups will be eminently successful in space, and others will fail, depending on how the individuals are constellated within them. Each space-station group, each lunar mining colony, and each planetary expedition will be, in a sense, a unique microcosm, an orbiting "ship of fools" or an experiment in survival. Each will have its own subtle mechanisms of control, punishment, reward, relaxation, and tension, and will define for itself what makes life human and worthwhile.

In a dangerous place like space, people will be asking these fundamental questions more often, and they will have to formulate adequate answers daily. Even now, the cosmonaut diaries are full of "life-review" answering the unspoken questions: Why did I come? What is the importance of my being here? Right now, the answer is usually the importance of the work they do. In the future, when space missions become longer, there might have to be more intrinsic or extrinsic ("I make a lot of money") rewards, a greater sense of the continuity of effort that brought the individual to the barren vacuum of space.

To every person, Carl Jung said, life poses a question. If a person does not answer that question for himself, then society will answer it for him. In space there isn't that luxury. Every existence uses precious life support, valuable resources, and is a vital link to all others; you cannot be superfluous, or have no purpose or meaning or justification to your existence. Every minute spent in space has to, in some way, be justified.

The mental landscape of space will be littered with pockholes of mild depression, group conflict, lack of privacy, and overwork. As we reach out beyond Earth's orbit to the Moon, Mars, deep space, the next threshold, people will once again tilt against loneliness, famine, ecological disaster, short supplies, death of children, debilitating or incurable disease. The price of exploring space will rise. The vast bulk of frontier is still very much ahead of us; we have barely scratched it or its rewards and terrors. Some might come home because they feel the price is too high, too much.

Why did pioneers uproot their entire lives, savagely cut every tie to the past and gamble their businesses and the lives of their families for rewards that were, at best, ill-defined and extremely speculative? The motivations had to go beyond wealth and social status, since the pioneers came from all the strata of society. The motivation was what might be termed "the call of the frontier."

"The end of an age is always touched with sadness," wrote Walter Prescott Webb, a historian of the American West who died at the dawn of the space age. "The people are going to miss the frontier more than words can express. For centuries they heard it call—listened to its promise and bet their lives and fortunes on its outcome. It calls no more."

But a frontier calls again, and many would heed the call if they could gate-crash NASA's elitist procedures. The nostalgia for the mythical West, which strides through our collective conscious in the form of movie and television "westerns," has collectively revived an excitement about future frontiers in the form of science fiction. These myths of new beginnings, authentic or not, are what people have always gambled their lives and fortunes on—and doubtless will do so again.

The demands of the profoundly alien form of space pioneering and homesteading will be a psychological challenge as great as the opening of the American West and the great voyages to the New World. But what is that price compared to the excitement and challenge of opening a continent, a world, or a universe?

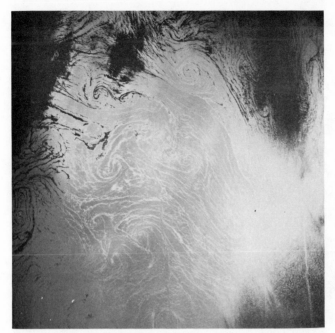

Bizarre swirls mark Mediterranean Sea, in this view taken by oceanographer Paul Scully-Power in late 1984. He took advantage of sun glint effects to record many such hitherto unappreciated ocean features. (*Courtesy NASA.*)

Windows on the space shuttles are much larger than those of any previously manned spacecraft, and the eyeball view is correspondingly better. A permanent space station must have viewports at least as good. Here, geologist Kathy Sullivan at the shuttle's front windows. (*Courtesy NASA.*)

An ice island tens of kilometers across, near the South Sandwich Islands off the coast of Antarctica, is photographed from Skylab. Orbital spacefarers can provide critical iceberg tracking, and can tell from the color of surrounding water just how fresh the icebergs are. (*Courtesy NASA.*)

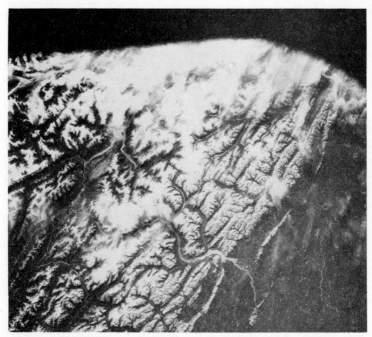

Snow-covered mountain range viewed from orbit. (*Courtesy NASA-HQ.*)

Only from space can our homeworld be viewed as a planet in its own right, with all the smallness, fragility, and vulnerability inconceivable before the space age. Our survival may well depend on how well we believe and act on this revolutionary concept of Earth. (*Courtesy NASA.*)

Both Soviet and American space stations have carried depressurization devices to stress heart muscles by pulling blood into the lower body, as gravity does on Earth. The Skylab device, right, was mounted on the floor (*courtesy NASA*); the Soviet device, left, produced less pressure, and was configured as portable trousers.

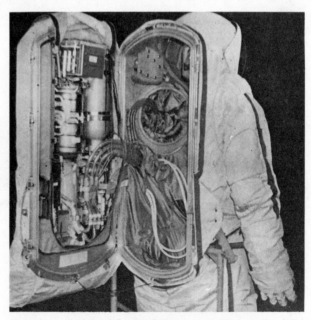

Ingenious design of Soviet Salyut spacesuit includes back door for cosmonaut's entry, and life support gear mounted on "door." These have been used repeatedly by different-sized Soviet spacefarers over multi-year suit lifetimes.

Russian spacewalker emerges from airlock hatch (comrade photographer visible in visor reflection).

Skylab spacewalker holds tight during film changeout operation with solar telescope. (*Courtesy NASA.*)

First man to command a space station, bearded Georgiy Dobrovolskiy, preparing to return to Earth. He had only hours to live.

Dobrovolskiy (center) with his crewmates Viktor Patsayev (left) and Vadim Volkov (right); all died on trip back to Earth when air leaked out of their spacecraft.

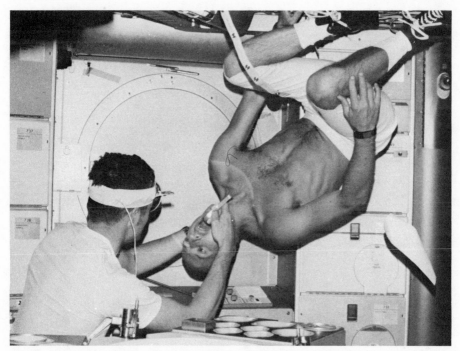

Dr. Joe Kerwin examines patient "Pete" Conrad in orbit; weightlessness affords more convenient posture for throat examination. (*Courtesy NASA.*)

In only a few years, spaceflight has developed a mature system of traditional behavior, some adapted from earthside norms and some wholly original. Here, returned cosmonauts Vladimir Lyakhov (left) and Aleksandr Aleksandrov (right) are greeted by earthsiders with old Russian custom of "bread and salt," a totem of hospitality. (*Courtesy NOVOSTI.*)

Most visions of space futures consist of vistas crowded with gadgets and gizmos of grossl
variegated configuration. (*Courtesy Boeing.*)

chapter eleven

GADGETS
— and —
GIZMOS

The establishment of long-life orbital stations with rotating crews is the main direction in the development of Soviet astronautics.—*Leonid Brezhnev (1982), Yuri Andropov (1983), chief cosmonaut Vladimir Shatalov (1984), and about two hundred others*

We can follow our dreams to distant stars, living and working in space for peaceful economic and scientific gain. Tonight, I am directing NASA to develop a permanently manned space station and to do it within a decade.—*Ronald Reagan, State of the Union Message, 1984*

Probably everybody has seen more than enough artists' conceptions of "space stations" and "space habitats." The illustrations demonstrate far more artistic license than hardware prophecies. A comparison with the past may be illuminating: merely glancing at half-century-old artist's conceptions of late-twentieth-century technology is enough to reduce most of us to fits of laughter. There is no reason to suspect that today's "concepts" will be greeted any differently by twenty-first-century spacefarers.

Nevertheless, future space vehicles will be an outgrowth of current technological developments. In the space-age equivalent to the chicken-and-the-egg priority question, today's space engineers are faced with an equally baffling dilemma: Which comes first, the form of a space station, or the function of a space station? Does one design a feasible platform in orbit, and then determine the best way to operate from it, or does one define all potential applications of such a space platform, and then build to suit?

Some space applications require lots of electrical power, and others require an unobstructed view of the Earth or of the Sun. Some require almost constant human intervention, while for others, a human heartbeat within a hundred feet is too much of a disturbance. Some require frequent changeout of large pieces of equipment: others merely need replacement film cartridges or filters or penny fuses.

Transportation issues are equally clouded. Will vehicles be designed to meet projected needs, and will those projections come true once the decade-long development is over? Or will we see a procrustean alteration of cargoes and missions to "make do" with transportation technologies we happen to possess at the moment?

Some cargoes need to be carried gently into space for a time, and then brought back. Most need only be dropped off in orbit. Some—a very precious minority—require special handling on their return. And others just need to be pushed from place to place in space. The requirements are so varied that no finite collection of capabilities will be optimal for each and every case. To meet these contradictory needs, designers muddle through by mixing and matching. And the results, although they generally manage to work, show it.

Space station architecture shows its pedigree: designed by the proverbial committee. Whether it is a "Space Operations Center" (Houston's favored approach in the early 1980s) or the "Pleiades Cluster" (from an academic study) or the "Power Tower" (which became the baseline for contract studies in 1984), each station layout contained implicit assumptions on station utilization and on functions that would not be adequately satisfied by any particular approach.

So swift is the evolution of these concepts, and the shifting priorities of desired applications for the station, that a far more productive approach is to look at known correlations between forms and functions of a wide variety of space platforms, so that whatever final configuration (or combination of configurations) is selected will make sense.

In 1983, a German satellite called SPAS (for "shuttle pallet satellite") was tossed out of the Challenger's cargo bay, then retrieved a few hours later. For NASA, it was a chance to practice maneuvering the big spaceship at close quarters with another satellite; for the Germans, it was the chance to conduct experiments with absolutely no disturbance from motions of crewmembers. The satellite floated absolutely free in space, taking full advantage of pure weightlessness. Once retrieved, it was returned to Earth where engineers removed the crystal samples created under such pure conditions.

Two years later, a desk-sized astronomy satellite called SPARTAN was released from a shuttle in much the same way as SPAS. This time, the bargain-basement satellite (which was loaded with instruments originally built to ride on sounding rockets) spent almost two days on its independent course. Following a preset program, it pointed itself first one way, then another, recording long exposure images without any disturbances. Finally, the shuttle swooped back from nearby, grappled it with the robot arm, and returned it to Earth. Project scientists recovered the data that had been recorded aboard on magnetic tape. Like SPAS, SPARTAN was battery powered and used gas jets to point itself.

At about the same time an even larger and longer-lived free-flier satellite was scheduled for return to Earth. Originally deployed in April 1984 during the Solar Max satellite repair mission, the LDEF (for "long duration exposure facility") had been the host for more than one hundred investigations that were arrayed like plates of armor across its cylindrical form. The LDEF was "passive," using the slight slope of near-Earth gravitational force to keep itself lined up, its long axis always pointed "up-down" (that is, toward Earth's center).

Sometime in 1987, the European Space Agency will be

putting a new payload on the space shuttle. It is called the "European Retrievable Carrier," or EURECA, and, like the other free-fliers before it, it too must be brought back to be useful. At deployment, a rocket engine will push it a hundred miles farther into space. There, it will unfold solar panels, and then conduct experiments and observations for up to a year before returning itself to a lower orbit, within range of a shuttle pickup.

These free-flier satellites shared a common trait, despite their sequentially greater size, electrical power, and lifetime: their experiments, which depended on absolutely undisturbed motion through space, collected and stored all data on board. Thus, retrieval was necessary to harvest the crop.

The next step in the evolution of unmanned free-flier satellites is a project called LEASECRAFT. Fairchild Industries is developing and funding this six-ton vehicle; as encouragement, NASA is offering several free rides into orbit, in exchange for which Fairchild will provide NASA with several berths at the facility. LEASECRAFT is a semipermanent, unmanned space platform, with standardized control, power, and pointing systems. Customers provide experiment modules that can be berthed to different sides of the platform. The modules can be developed much more cheaply since they need not concern themselves with "housekeeping" functions (instead, they pay rent and utilities to Fairchild). They would be brought up and retrieved by shuttle missions, much like earlier free-fliers, and up to 33,000 pounds of modules (a limit set by the maximum weight the shuttle could safely bring back to Earth) could be bolted onto the "bus." The spacecraft's design is to be based on the "bus" portion of NASA's Solar Maximum Satellite (repaired in orbit by space shuttle astronauts in early 1984) and Landsat (one of which is a target for an astronaut repair mission in 1986).

Fairchild plans to build and launch two LEASECRAFT platforms, and could have a fleet of ten in space within five years. With this concept the two orbital functions of productive activity and basic support have been separated, and the basic support structure remains in space indefinitely (up to twelve years, according to plans). Shuttle visits would service the platform with rocket fuel and other expendables, as needed.

Sometimes it may be too expensive to retrieve an entire experiment module, when only a sample cannister or a film cartridge needs to be replaced. Such servicing must still be done by shuttle flights, using expensive spacewalking techniques. This can be done with the LEASECRAFT, but it must be done with such unmanned scientific orbital observatories as the "Space Telescope." However, a more convenient (and hence cheaper, in terms of servicing charges from NASA) system is being developed.

Space Industries Incorporated in Houston wants to build a LEASECRAFT-type platform with a major difference: It will have a pressurized interior, and the attached specialized modules will have hatches through which "shirtsleeved" (as opposed to "spacesuited") astronauts inside the platform's hull have access to all serviceable hardware. The platform, an "almost-space-station," would have no true life support, but would depend on fresh air from the shuttle's airlock, which would be docked to a port at one end of the platform. Astronauts would not spend more than a few hours aboard on any one service call anyhow.

These visits need not be very expensive, since they can be scheduled to be incidental to the shuttle's main mission of launching a series of satellites. The add-on cost of the rendezvous and servicing would probably be no more than a tenth of the cost of a dedicated launching, not orbitally exorbitant at all!

The evolutionary sweep of these sequential types of spacecraft is evident. Step by step, it is approaching—from the direction of a longer-term orbital platform—the general concept of a space station of the future.

There have been space stations in the recent past. The first—the Soviet Salyut and the American Skylab—shared some characteristics with each other, but not with their descendants.

Both were launched aboard expendable, unmanned boosters. Both carried most of their scientific gear and life-support supplies aboard. Their available electrical power was not extensive, Salyut providing about three kilowatts and Skylab about nine. At the end of their missions, usually within a year, they were abandoned in space.

The Soviets chose an evolutionary improvement approach, and extended the lifetime and resuppliability of their platforms, recently making some small "quantum jumps" in capabilities. But the U.S. has chosen a "great space leap forward" that leaves Skylab and expected Soviet counterparts far behind. But for the moment, a little more mixing and matching of gadgets.

Spacelab is a caboose-sized pressurized module used to conduct scientific activities for up to ten days in orbit, while attached to the inside of the shuttle's payload bay. Spacelab is loaded with scientific apparatus, but the shuttle's onboard equipment provides all pointing, life support, electrical power, and communications links with Earth.

At one time, it was promised that the shuttle could support orbital flights of up to thirty days. With Spacelab modules inside, missions could have been perfect scientific sorties with specialized apparatus too experimental and heavy for permanent installation aboard a space platform. But that promise faded, even though proposals for a specially modified "Long Duration Orbiter" (capable of flights of forty-five to sixty days, or even more) have been around for years. NASA has various reasons, both good and bad, for skipping this stage: first, with so few shuttles actually built, taking one out of service for such missions would squeeze out customers for satellite launchings; second, the availability of such a capability, together with unmanned space platforms, would appear to undermine the argument for a permanently manned space station.

It has not escaped the notice of European space scientists that Spacelab modules might be able to look elsewhere for such on-orbit utilities. One suggestion is to hook a specially modified Spacelab module to some platform such as EURECA or LEASECRAFT (life support and communications would have to be developed separately). This would produce an embryonic space station, for the most part merely from modifications to preexisting components.

Alternately, a Spacelab module could be hooked to another preexisting platform, the Soviet Salyut. Getting Spacelab into orbit and over to the Salyut could be diplomatically delicate, since only a space shuttle could carry it. But the Spacelab would add tremendous scientific capability to the Soviet platform, while deriving all its utilities from it. In engineering terms, however, such a lashup would not work because the basic Salyut module is woefully underpowered: aboard the space shuttle, Spacelab is supplied with up to ten kilowatts of electric current, while the total Salyut output is under four kilowatts for everything.

To the Soviets, the advantage of adding some sort of specialized modules to their basic core platform seemed attractive. In the early 1970s, they decided they needed dedicated scientific modules (sort of mini-Spacelabs), more electrical-power generators (solar panels), and larger fuel capacity on orbit.

According to a mid-1984 statement by Vladimir Kotelnikov, vice-president of the Soviet Academy of Sciences, "Soviet cosmonautics is confidently and consistently promoting the creation of intricate orbital manned complexes that can be reconstructed and refitted even in the process of flight according to the nature of the tasks being performed." On Cosmonaut Day (April 12) 1984, space expert Mikhail Chernyshov had written that "The presence of cosmonauts is absolutely necessary for certain types of work, and, conversely, their presence is undesirable in other cases. It is already clear to the specialists that a harmonious compromise in meeting all of the contradictory requirements can be achieved only on the basis of multiple-module stations, although multiple-modularity does not mean that all modules must necessarily be linked rigidly into a single cluster." Chief cosmonaut Vladimir Shatalov had, earlier that same year, given details: "The future of Soviet astronautics [will be] space complexes composed of separate units. . . . The units will contain research equipment, laboratories, and workshops. I think that some units will remain aloft for some time in an autonomous flight without communication with the station, according to a set program. They will be docked then to the basic station for repair, exchange of equipment and supplies of medical preparations, extra-pure materials, and so forth." Chernyshov had further predicted an evolution in the appearance of the central space station: "The central nucleus of the station itself will no longer look like a scientific laboratory, filled to the limit with various kinds of apparatus. More likely, it will be a kind of living area with the most comfortable conditions that are possible in space."

The orbital tests with add-on modules such as Kosmos-1443 in 1983 are advertised as steps in that direction. Commentator Pyotr Pelekhov declared that year that "Future space equipment will consist of an extensive set of specialized modules, that is, functional units. . . . The development of modules—laboratories, workshops, observatories, and so forth—is being proposed on the basis of the Kosmos-1443 module." Vladimir Lyakhov, one of the cosmonauts who had been aboard the module when it was docked to Salyut-7, praised its value: "Such a marked increase in working area improves the crew's psychological and physiological state, and makes it possible to install a continuously-[deployed] shower and a workshop from which shavings or waste products no longer get into the station's living quarters." Lyakhov was referring to the trouble it took to unfold, use, dry out, and refold the collapsible shower stall in Salyut-7; it was obviously one of his least favorite features of the space station.

The problem for the Soviets remains that such modules must be launched aboard expendable "Proton" boosters until the early 1990s when their large Columbia-class "space shuttleski" becomes operational.

For the American space station, a similar direction is forecast: six to eight specialized modules will be strung together to form the permanent orbital outpost. Current thinking is that there will be two habitat modules, two laboratory modules, one supply module, and a power and structural complex brought up on several initial shuttle missions. There will also be free-fliers serviced periodically, either by their docking to the station or by astronauts visiting them from the station.

The station will have one docking port for the visiting space shuttle, at least one personnel airlock, and more than a dozen attach points for exterior experiments. In fact, these "piers" will look like skeletons of the shuttle cargo bay, so that equipment designed originally to operate in the bay on short missions can readily be installed aboard the station.

One uniquely American function of the station is to serve as a fuel depot. Most of the weight carried into low Earth orbit is fuel, not hardware, and once in orbit it is quite literally worth many times its weight in gold. Since the shuttle will service the space station, and most shuttle launchings will not carry full cargo loads, any extra capacity can be filled with fuel for ballast (there are thousands of pounds of "wasted fuel" in the shuttle's lines when it jettisons its belly tank—nowadays, that precious commodity is blown overboard every mission). This fuel would be accumulated aboard the station, and allocated to users as needed.

The theoretically optimum transportation system between orbit and Earth's surface is a complex of elevators, dumbwaiters, and pneumatic tubes. Different sorts of cargoes, from freight pallets to express mail packets, from human beings to emergency medicines, could be transshipped between Earth and space economically, on short notice, and harmlessly. Centuries from now, such wizardry may become technologically feasible. For the foreseeable future Earth/space/Earth transportation will remain the domain of belchfire-mode rocketships and fireballing reentry capsules.

First there were the "capsules" such as Vostok, Mercury, Apollo, and Soyuz. They were roughly cone-shaped, to provide survivable braking when they hit the atmosphere on their way home; in space, they usually had "service modules" of extra equipment (air, electrical power units, small rockets) hooked on behind, which were jettisoned just prior to hitting the atmosphere. They were launched atop throwaway boosters, originally designed (or at least based on) military missiles.

They weighed a few tons, and could deliver a few people and a few hundred pounds of discretionary cargo into orbit and then back to Earth. In orbit, they could perform brief missions on their own, or could hook up to larger modules, the early space stations, for a few months.

The Soviets optimized their Soyuz design to serve as a pure cargo carrier in an unmanned, robot-controlled mode. Once the crew compartment was removed, the same airframe/booster that could take three people and two hundred pounds of payload into orbit and back became capable of carrying a full load of five thousand pounds into orbit (with no return—the supplies were for people already there).

In 1982, Russian engineers added a congenial touch. Usually the capsule's cargo hold had to be loaded and sealed several weeks in advance. But a special hatch was built into the compartment's side to allow a small volume of urgent cargo to be loaded on the launch pad only hours before blast-off. Traditionally, this came to be used for sending up fresh foodstuffs such as bread, milk, and berries, as well as recent newspapers and the very latest letters from home.

Although strictly speaking the Progress (as the modified craft was called) was only a one-way transport, it rarely left the Soviet space stations empty. Cosmonauts regularly loaded them with massive volumes of trash, soiled clothing, waste

paper, toilet filters, expended air cartridges, empty food containers—and then jettisoned the vehicle, aimed for fiery destruction over the South Pacific Ocean.

The American space shuttle provides a tremendous (some
would say excessive) capability beyond this level: It can carry
up to ten people and thirty tons of payload into orbit and
back. Since most of the cargo on planned missions of the
1980s involve only one-way flights, some observers have suggested that an unmanned shuttle-based booster system optimized only for "up-cargo" could carry three or four times
as much cargo, at cheaper freight rates, than the kind NASA
opted to build.

While the criticism may be technically valid, it ignores the
key ingredient in planning any major space program—political feasibility. The shuttle, even potentially more expensive,
was (probably rightly) judged much more affordable in terms
of federal budgets. The same criticism has been leveled at the
federal commitment to a full-fledged space station, again with
much the same technical validity and with the same naïveté.

However, it would be wrong to think of Columbia, Challenger, Discovery, and Atlantis as the only way to build space
shuttles. Designers in the U.S.S.R., France, Japan, and the
U.S. Air Force have been working on smaller winged orbital
ferry craft, to provide many capabilities for which the NASA
space planes are not fit. For example, a small, two-man shuttle
craft could bring urgent cargoes or personnel to a space station, or perform rapid sorties into orbit for rescue, repair,
servicing, or military functions.

Meanwhile, the shuttle could take advantage of its tremendous upgradability—the built-in expandable features that (like
the lower deck on the George Washington Bridge) seemed
like wasteful luxuries when first installed but could eventually turn out to be extremely farsighted. One exciting example is a passenger-variant design, in which the shuttle
carries a personnel-transfer module in its payload bay. Designs exist for carrying up to seventy-four people into orbit,

but that would be feasible only if there were some destination such as a major construction project awaiting them. Another design, put forward by Planet Earth Productions in late 1984, would have a fifty-four-passenger module and a viewing lounge for space tourism—$2 million a ticket.

Both capsules and standard shuttles take off and land with about the same cargoes. This conceptual habit can be broken by considering systems designed for one way or the other. The Soviet Progress freighter-tanker is such a "one-way" access device. Getting back from orbit has some interesting angles.

Unmanned capsules did it first, dropped by prototype military reconnaissance satellites. Later capsules carried Soviet samples of moon dirt, biological specimens, or film cannisters from American, Chinese, and Soviet photo-reconnaissance satellites. Cargo rarely exceeded a hundred pounds in weight; the whole recovery capsule may weigh upwards of half a ton.

If aimed accurately enough, such small capsules can be dropped right onto some recovery reservation in the semi-desert of central Asia (on either side of the Sino-Soviet frontier) or the western United States. American space engineers used to prefer ocean landings, both to help cushion the impact and because large naval forces could be rapidly deployed to any preselected region on short notice.

But the most common way of recovering American orbit-to-Earth capsules was pioneered a quarter-century ago, and has remained largely out of the public eye ever since: air snatch. Since it is the Air Force doing the snatching, and superspooky spy photographs are the capsules' cargo, such a low-profile mode is understandable. As each capsule descends by parachute near Hawaii (on at least monthly intervals), large transport planes trailing special "trapeze" hooks snag the shroud lines, and haul the cargo in. The film can be flown to processing centers, and be on the analysts' desks within hours.

For the Soviets, the orbit-to-Earth leg was a real bottleneck well into the mid-1980s. They attempted to solve it by including one special-purpose "down-cargo" capsule with each specialized add-on module sent up for attachment to the core Salyut space station. These capsules had an advertised capability of half a ton of payload weight brought back to Earth. The design had been flight-tested as part of the short-lived Russian manned military reconnaissance space stations of the mid-1970s (evidently scrapped after losing a "fly-off" competition with improved unmanned space spying systems), but it took the better part of a decade to redesign the whole module for use as an add-on to the basic Salyut. It was not until mid-1983 that cosmonauts actually utilized this "down-cargo" scheme, and it was another several years before they were able to do so a second time.

NASA's plans for return cargo from orbit center on retrieving satellites (using the robot arm and jet-packing spacewalkers) and on exchanging "pantry modules" with the planned space station. Under the latter mode, resupply visits would occur every few months, and would involve bringing up a fresh-stocked supply module that would actually be docked wholesale to the side of the station; the used-up pantry, also loaded with scientific and industrial products as well as dozens of trash bags, would be retrieved. Additional spoilable cargo (both up and down) could be stowed in the two-and-a-half-story shuttle cabin.

The existence of this wide variety of down-cargo systems demonstrates the available options for space-station planners. How much weight and volume will be needed, and how frequently on a routine basis? What kinds of cargo will need urgent short-notice transport? What temperatures and pressures must be maintained, and what acceleration profile will the cargo have to endure (the shuttle brings home the bacon at a gentle one-and-a-half g's, while standard unmanned capsules undergo six to eight g's—too much for some spacemade products to endure)? How fast does the returned material need to be delivered to users, and under what conditions of biological isolation or military security?

Different requirements have thus already led to the existence or development of several different systems, both for up-cargo and down-cargo capabilities. Several gaps can be seen in this array of capabilities, however.

First of all, once a space station is fully established there are numerous scenarios where a small cargo (perhaps a replacement part, perhaps some special medicine) will be urgently needed but will not justify an entire space shuttle mission. In such a case, a small minishuttle or a fast-response unmanned system could prove cost effective.

Second, as already mentioned in Chapter 9, there is a particular type of urgent down-cargo that could need transportation: human beings. Whether for medical, psychiatric, or disciplinary reasons, individual crewmembers may require evacuation long before the planned arrival of the next shuttle. Even an urgent, ad-hoc shuttle visit, such as for a medical emergency, could take weeks to schedule.

Additionally, there may be situations in which the station becomes no longer habitable. Most of the time, there will be no shuttle craft docked at the outpost, so the only currently planned option is to hang tough in an emergency shelter— for weeks if need be—or die.

Rescue is problematical at best. NASA's plans for putting the station at an orbital inclination of twenty-eight degrees means that Mother Nature herself rules out the possibility of any Soviet rescue craft (such as is allegedly on standby whenever cosmonauts are aboard the Salyut) reaching that orbit from their launch site at forty-six degrees north latitude (the plane change would require half again as much velocity increase as it took to get into orbit). Even the Vandenberg shuttle launch site is probably too far north to support a rescue launch. This leaves at best three (maybe only two, if one is down for maintenance or cannibalization, as seems more and more likely) vehicles in Florida, and with launchings about two months apart, it would be ten or twenty days before the very next in line vehicle could be made ready.

Such considerations have prompted some space engineers

to advocate the development of orbital bail-out systems, by which spacefarers could return to Earth on their own without recourse to a rescue mission from Earth. One suggestion is a capsule based on the old Apollo command module design, serving under normal conditions as a command post, a store-room, a bedroom, or even a bathroom; in an emergency, several crewmembers could squeeze inside, cast off from the station, and head for an emergency landing on the next available landmass. More stripped-down concepts involve the development of an even smaller capsule to be routinely used for returning urgent cargoes between shuttle visits, which in an emergency could accommodate a crammed-in human passenger. Most basic of all is an old proposal from Apollo days, in which a spacesuited astronaut used an inflatable fast-setting-foam heatshield and a chest-mounted, eyeball-aimed, solid-fuel retrorocket to get himself back into the atmosphere; once below about twenty thousand feet, the evacuee would cut himself free from the cocoon and pop a personal parachute for a "routine" descent.

The feasibility and desirability of such systems remain to be established, but there is little doubt they will someday be developed in at least one spacefaring nation. The personal impressions of the first spacefarer actually to "jump" from a spacecraft in orbit are going to be sensational.

From a point in space, there are many other plausible destinations besides just a return to Earth. There are other worlds, and there are an infinity of other points in space. For departures into interplanetary space, the station could best serve as an assembly and refueling point for the spacecraft, particularly if, as is hoped for the American system, the station can scavenge excess fuel from routine shuttle missions and thus accumulate a considerable and valuable reserve at little operational cost except patience.

Many of the free-fliers associated with future orbital outposts will be in orbits considerably higher. They cannot be reached by shuttle (or Soyuz) missions, so some sort of "space tug" (the Soviets use the term "space crane" as well as "tug") is needed.

NASA's plans are incarnated in a spacecraft called the "Orbital Maneuvering Vehicle," or OMV. Since the shuttle ticket office charges by the linear foot of payload bay, the OMV is to look like a Christmas fruitcake tin, a stubby cylinder fifteen feet in diameter and only three or four feet thick. It would fly under remote control to the unmanned spacecraft of interest, dock with it, and push it back to the station for servicing; afterward it would reverse the process.

Ultimately, a manned module could be attached up front. But in that case the astronauts would no longer be able to land directly; in case of trouble they would have to get themselves back to the station or get rescued. Already, such a "nonreturn" manned spacecraft has flown, once, in Earth orbit: in early 1969, a manned test of the Apollo "Lunar Module" involved flying a hundred miles away from the command module (and their tickets home). Lunar Module flight controllers in Houston had their own cocky motto: "Heat Shields Are For Sissies." Sometime in the 1990s such cockiness and confidence may be called for again!

When a spacefarer gets back from a months-long tour of duty on a space station, he or she is not physically up to climbing ladders and stairs to get out of the return spacecraft. The Soviets trundle a long-legged platform over to their landed Soyuz capsules, which has several access ladders and a working platform from which rescuers can reach down into the Soyuz and extract the limp spacefarers. They descend from the platform on a slide, held carefully all the way by the rescuers. Once on the ground, they can be carried, or they can take a few steps, to divan chairs, which can be lifted onto the recovery helicopter.

For several more years, returning American space shuttle crews will still be able to bounce briskly down the steps of the airline-style mobile staircase rolled up to their side hatch soon after landing. But the first team of Americans that comes back from several months in space in 1992 or 1993 will certainly not be allowed to attempt a pedestrian negotiation of those stairs. Some entirely new kind of structure is needed to get weak and shaky spacefarers out the shuttle hatch and down to the ground. It may have an elevator, or a set of stretchers or easy chairs on a forklift (like the one that greeted the Skylab astronauts), or even wheelchair ramps. Getting an invalid of any kind out through the round shuttle hatch is going to be quite a challenge for the gadget and gizmo builders. In fact, it's likely that the entire seating arrangement on the shuttle middeck will be rebuilt so the spacefarers can lie prone during the entry deceleration forces, rather than sit up and let the blood drain from their brains.

The importance of all these pieces of space hardware must be measured primarily by what they can do for people's needs and goals. Sometimes the perceived need comes first, and a gadget must be designed and built, or adapted from something already found in the inventory. At other times, the existence of a gadget, developed and paid for as part of an entirely different program, can give spaceflight engineers new ideas for novel and unpredicted utilizations of it. (Space engineering experience has shown, however, that it can often be more expensive to modify and validate an existing gadget than to develop one from scratch—the Skylab's "orbital workshop," a modified fuel tank, is a prime example.)

In any case, the gadgets and gizmos of spaceflight are so impressive, so expensive, so sexy, so easy to visualize and present in the superficial overviews characteristic of the general news media, that they can all too easily distract attention from the truly important theme, which is what the hardware is for. It is for enabling and facilitating the human homesteading of outer space, and if that concept isn't exciting enough, no mere piece of equipment should be either.

chapter twelve

BATHROOMS
— and —
OTHER PLACES
— of the —
HEART

There's the toilet and the refrigerator—what more is there in life?—
John Belushi, "Continental Divide"

What makes a place livable? Toilets that work. None seems to in space. Every toilet on Earth is gravity-dependent, and the people who use them have gravity-dependent toilet habits. If there are any space wars, the first battlefield will be the toilet.

Early space travelers had to use a bag with a sticky mouth that adhered to the rump, which had little pockets, so fingers could do what gravity couldn't—direct a stubborn "bolus" (as it is called in medical literature) into the bag. On one of his flights, Wally Schirra "held it" for eleven days rather than use this contraption (and it showed in his general attitude).

Urination in the early spacesuits was a bit easier; it involved a collection bag, a hose, and condoms of different sizes. Legend has it that the astronauts always chose the largest condoms, machismo being what it is, and caused a few soppy messes before NASA did away with the designations "small," "medium," and "large."

On Skylab, the toilet seat was mounted on a wall. People still defecated into a bag, but it was a porous one, and a suction device at the bottom of the toilet helped the bolus along, and carried away the odors too. To please the medics, the patient Skylab astronauts had to label each sample, weigh it, place it in a processor for vacuum drying, and stow it for metabolic analysis on return to Earth. The urinal involved a hose and a universal funnel (the latter very democratically did away with size discrimination).

The Skylab toilet worked. The only trouble was that urine collection by the condom or funnel method was clearly impractical for females. The shuttle's answer to the male-only Apollo paraphernalia was the million-dollar coed toilet. Affectionately referred to as the "Slinger," it was shaped pretty much like an Earth toilet, with some important modifications, and was located on what we would all recognize as the floor. It had foot restraints and a seat belt to keep the person from floating away. One had to sit in such a way as to create a seal between derrière and toilet seat, or risk letting urine or feces escape and seek some other place of rest. There was a suction device at the bottom of the bowl to pull the urine and feces down. When finished, the person would close the toilet lid and flush. However, in flushing, the space toilet activated a set of tines that shredded the feces and deposited it on the toilet bowl wall (some designer with a mordant sense of humor must have thought of this—for once, the shit literally hit the fan). This unsightly mess was then vacuum dried to the side of the toilet; layer upon layer was deposited on the bowl wall during the entire flight. It wasn't cleaned until after the flight, when some unfortunate person working for some NASA contractor had to literally chip the stuff off. Needless to say, the shuttle toilet was designed only for short spaceflights.

One of the greatest sight gags from a science fiction movie

was in *2001—A Space Odyssey*, when Heywood Floyd, the Moon-bound colorless bureaucrat, has to read a two-page, single-spaced instruction sheet for using the space toilet. The shuttle has a similar set of instructions on the wall beside the toilet.

This toilet worked only in theory. In practice, it had problems. On the first shuttle trip, a component shook loose on launch and blocked the vacuum suction. On a subsequent flight someone tried to flush a vomit bag down, and clogged the tines. The designers confidently and cheerfully maintained that the toilet would work as advertised if the astronauts were sufficiently "potty trained." Most of the astronauts have graduate degrees, many in engineering; they are capable of using the most sophisticated, technologically advanced equipment in the world. But NASA alone believed this outlandish notion, and followed it up with more potty training. Only after numerous, consistent failures did someone in authority realize that the toilet was badly designed and difficult to use.

One had only to speak candidly with the users. Maintaining a seal with the toilet seat was difficult because the designers had disregarded the fact that human bottoms, like human beings, are not created equal or alike. The slinger's method of feces disposal was unpleasant, if not unsanitary. Some other parts also clogged with fecal buildup, and had to be cleaned after each use. Crews complained that it was a mess to change the urinal filter, a process that liberated a significant amount of urine each time. Though the astronauts have never discussed it, who would really want to use a toilet whose unsightly appearance reminds one of all who have gone before? The foot restraints and the seat belt were not really equal to the task of keeping someone down on the toilet. Some astronauts favored a thigh restraint, in the form of a bar that could be swung across the lap—one could be held down and have someplace to rest one's reading material.

Because the slinger was designed for use on short shuttle flights, it is not the ultimate space process of elimination, in spite of its price tag. Not only does the space-station toilet (maybe even a two-holer) have to work, but it has to work quickly, cleanly, and efficiently to accommodate eight to twelve people forever. New designs are already on the drawingboard. Some have suggested placing the toilet beside a window so one could forget the indignities and enjoy the view; at least there, one could sit secure in the knowledge that nobody would look in and molest one's privacy.

Out of an almost endearing sense of delicacy, the Russians don't really talk much about their toilet facilities. In fact, their prudery is so profound that their news media even ignore the frequent toilet problems on space-shuttle flights, although they regularly glory in reporting all other American problems in excruciating detail.

However, there has recently been a leak in the excretory curtain. Cosmonaut Aleksandr Ivanchenkov discussed the Salyut toilet design with Russian correspondent V. Sudakov, for the English-language edition of the magazine Soviet Union: "The compartment between the working compartment and the passageway to the docking unit is sectioned off by a rubber curtain with a zipper. This is where the sanitation facility is located. Everything is as close as possible to a toilet on Earth. The bowl is standard with the exception of the urine dispenser, which is separate and attached to a vacuum. A plastic insert with a bottom filter is put into the bowl. Upon use, a rubber valve momentarily seals it, and the feces are packed away in hermetical rubber bags, which are put into a plastic container and ejected into space."

The shuttle toilets have been the source of vast amusement, but after removal of the fecal shredder, things seemed to have settled down. But more problems may be looming. It is only a matter of time before some perverted doctor asks for an anaerobic stool sample. That would require defecating through a slot into an airtight, nitrogen-filled container. With a perfectly straight face, he may ask for not just one, but a whole series of samples from several individuals, both in space and on Earth (as a control group). Anaerobic stool samples, he would claim (probably correctly), will undoubtedly shed a great deal of light on gastrointestinal function, microbial life or lack of it in the bowel, and various metabolic processes. But humankind may have to do without this particular tidbit of enlightenment, just as it has so many others that violate human dignity and fly in the face of feasibility.

As it stands now, too much already assaults everyday human dignity in space. Apart from the myriad difficulties of going to the bathroom is the problem of smells. There is, of course, the not widely mentioned matter of unavoidably exuberant flatulence, one of the aromatic charms of weightlessness. Because people cannot burp, they must pass their gas elsewhere. Happily, the sense of smell is somewhat degraded in space due to excess accumulation of fluid in the head, like a bad head cold.

The other rarely mentioned hygiene indignity is the inability to take a really good shower. Water is in short supply in space. If any of it floats freely, it forms droplets, and scatters everywhere. On the shuttle, people have to wash up with wet-wipes and towels, which does not make for a really clean clean.

On the Salyut, cosmonauts rely on wet-wipes ("They are always cold," one remarked bitterly, although they can be moistened with hot water), and are allowed to use the shower ("if you can call it that," sneered another) every two or three weeks.

As described by Ivanchenkov, the Salyut shower was very similar in design to the Skylab system. It consists of a flexible, accordion-folded plastic cylinder with two five-liter water tanks, which unfolds from the spacecraft's ceiling. Before showering, "We turn on an electric water heater and let down a polyethylene cylinder [the shower] from the disk to the floor. Fastened to the bottom of this chamber are rubber slippers for the feet, to prevent you from floating upwards. We then get undressed, unzip the cylinder's water-tight zipper, climb into the shower and seal ourselves in the same manner. Above our head is the shower head and cellophane bags containing napkins and a towel. Before turning on the water we place in our mouth a mouthpiece attached to a hose leading outside the chamber and we block up our noses using clamps. Next we open the bag containing a small towel that has been saturated with soap and turn on the water, which comes out in a fine, needle-like spray. The wet mist is sucked out of the chamber through small holes in the floor by a vacuum system and discharged into the daily waste container. After a fifteen-minute-long shower, we wipe ourselves and the walls of the chamber dry. Finally, we lift the cylinder back in place on the ceiling."

The shower's one major flaw was that it was hard to dry after use. If the walls were not wiped completely, the shower would mildew and stink in a very short time. Finally one ingenious engineer devised a hula-hoop affair that the cosmonauts could use as a circular squeegee. They were overjoyed when it worked well.

But even that wasn't like a real bath. Aboard Salyut for over six months, Aleksandrov once dreamed of taking a bath, a real Russian bath complete with sauna, cold shower, and rubdown. That deep down clean feeling was something that every long-term space voyager misses.

Garbage piles up in space like it does on a New York street during a sanitation workers' strike. After a few days, it even begins to stink, too. The shuttle astronauts often have so much (over half the food weight on the shuttle is containers and wrappers) that they can't stow it for landing; instead they strap the bags down on the middeck floor and hope for the best.

The Soviets, whose orbital anti-litter habits apparently aren't as highly developed as Americans' from countless "Don't Be A Litterbug" campaigns, regularly dump everything (garbage, excrement, dirty underwear—*everything*) overboard. American satellite trackers who watch these little packages on radar affectionately refer to them as "honey buckets," as they spiral down to a fiery disintegration. The Russians refer to them as "little sputniks." Jettisoning them is one practical solution to going months without a garbage pickup. Also, every other month or so, a Progress robot freighter links up. Once unloaded, its cargo hold becomes an orbital dumpster too. At the end of its mission, the robot ship plunges back over the South Pacific. Everything is burned up in that giant cosmic incinerator called Earth's atmosphere.

Food is another issue that gets people where they live. Past astronauts have had to squeeze chicken and dumplings or bacon and eggs out of a tube. Later, they used hot-water squirters to heat up and hydrate dehydrated food. Now, NASA food technologists tell us that they are eating off-the-shelf food (one is tempted to ask, which shelf?). On early shuttle flights, astronauts complained that the water had bubbles in it, and wasn't hot. The water dispenser's needles became clogged and sometimes even failed. Meals took too long to prepare (at least forty minutes), and food containers would float out of their places on the tray due to insufficient anchoring. Some of these problems were solved in later flights. The shuttle is sometimes equipped with a pseudoluxurious galley containing a convection oven to heat rehydrated dinners, although sometimes the spacefarers still have to resort to more primitive heating trays when a commercial customer needs the middeck space.

Most of the astronauts have pronounced the food "pretty good" (especially the barbecue beef), although some can't resist a joke or two. When John Young and Bob Crippen landed after their STS-1 flight, they were told to remain in the shuttle until the technicians were certain that all noxious gases in the area were dispelled. "That's just the frankfurters we had for lunch," Young quipped.

Food technologists feel that most future space food will be heavily processed—irradiated, freeze-dried, dehydrated, possibly extruded. Food as we know it on Earth—crisp salads, fresh fruit, juicy meat—would be passé, a rarity, a delicacy.

It's bulky, generates a great deal of refuse, and always poses a storage problem.

Consider the apple. You don't eat the core or the stem, and many people don't eat the skin. It doesn't stay fresh for long, maybe a few weeks if stored in the refrigerator. While it might be delicious fresh, juicy, and tart, it could be cored, peeled, and dried, and then stored indefinitely. A whole barrel of dried apples could fit into a relatively small volume, and there would be no refuse, no waste. Of course, there wouldn't be that nice juicy apple flavor and crispiness.

If that weren't enough, there are those at NASA who believe in better eating through chemistry. "Let them eat agar-agar or polymer compounds!" was the thrust of a scientific paper given by two NASA scientists, Dr. Paul Rambaut and Frank Samonski in 1979. They believed that synthetic food, stored nutrients, and microorganisms would also be good food sources, given artificial palatability. And who needs variety? "Experimental evidence shows that individuals can be kept on a single nutrient source for many months without suffering ill effects," said Rambaut. His Soviet soulmate once had a test subject eat chlorella crackers for a month. There is no evidence this test was ever repeated.

Potential space colonists need not fear from these grim gustatory predictions. Futurists in the 1950s said by 1984 we'd be eating dinner as a pill, and flying helicopters to work. We've heard all these kinds of stories before.

Companies who send people to remote places on Earth can certainly tell spaceflight planners a thing or two. In a sensory-deprived locale, food takes on importance far beyond what it would be in ordinary circumstances. This is even true of people who are usually indifferent to food. People use food to get rid of boredom, frustration, anger, tension, and as a reward or celebration. In sensory-deprived places where there is plenty of undispelled tension, lots of pressure, no escape, and few physical pleasures, food's psychological function as a pacifier and tension releaser intensifies.

In space, the question of food is further complicated by a

degraded sense of smell and taste. Even under the best circumstances, space food quickly bores people. Astronaut Jack Lousma said that foods kept their flavor, but all food took on an aura of sameness. Eventually, everything had the same unappetizing smell and flavor, and they wanted something different.

Foods that people favor on the ground they do not necessarily favor in space. The cosmonauts on early Salyut missions went through their six-month condiment supply—especially horseradish, garlic, and other strong flavorings—in only a few weeks. American astronauts take a healthy supply of taco sauce and mustard, along with mayonnaise, catsup, and other standard flavorings.

The Russian cosmonauts typically go through their supply of soup first. Soup is satisfying, earthlike, and familiar. Borshch reminds them of home. Skylab astronauts asked for more filet mignon and ice cream, because they taste good, and possibly because they are treats, foods associated with reward and celebration. Both Russians and Americans wanted foods that had psychological comfort as well as gustatory pleasure.

The earliest forms of space etiquette and customs have to do with food. On Skylab, nobody floated over the dinner table to get to his place, even though it was easy to do so; instead, they squeezed in past people already at the table, as they would on Earth. On Salyut, the Russians greet their temporary visitors by offering them bread and salt, an ancient Russian tradition of greeting and sending off travelers. It involved elaborate preparation: "We prepared a large loaf of bread from about 50 small rolls," recalled Ryumin. "To do this, we had to prepare a base to which the small loaves of bread were sewn. We sewed a large loaf to a towel and the towel to the metal window shades [which undoubtedly served as a platter]. Above, we sewed a partition shade into which we put three tablets of cooking salt. This turned out to be an excellent bread-salt welcome."

Nasty episodes can occur as a result of the use and misuse

of food in an isolated, remote station. One such disaster occurred years ago when a chef served up a supper of authentic quiche Lorraine to a construction gang on the North Slope. Their idea of dinner was half a buffalo. On the other hand, one early, very successful Antarctic scientific group came back an average of twenty pounds heavier because their chef was a natural nurturer who took pleasure and gratification in his job.

Alcohol, on the other hand, has always been a trouble-maker. It unleashes suppressions, and there's plenty that people in isolation suppress. Many violent episodes can be linked to alcohol. In one case, a young man became very drunk after learning of his father's death and went on a rampage. Nonetheless, alcohol, under controlled circumstances, is a psychological booster. On some supertankers, there is something called "pour-out," an established, controlled routine where people can have their drinks at a certain time every day. It's a little like "happy hour," a time of conversation and relaxation, although, unlike "happy hour," it never degrades into violence and drunkenness.

There are some in the Soviet space program who wouldn't be averse to allowing the cosmonauts to carry a modest amount of vodka aboard (if they haven't already done so). Americans have enforced a "no booze" rule ever since the beginning of the space program.

"No sex" is going to be harder to enforce. The whole question is being handled, or rather ignored, as a matter of interior design. Bedrooms on space stations will be cubbyholes supposedly too small for two people. In spite of the nearly quarter-century-old sexual revolution, and the daily bombardment of sexual material from television, magazine ads, and store window displays, spacecraft designers see bedrooms as places where people sleep, not where people recreate. Worse yet, if bedrooms are located along the exterior wall of the space station (where most designs put them), one has Newton's third law (every action has a reaction) to contend with: *every little motion imparted on the wall will jiggle the ship and register on the instruments.* This more or less public display of affection may be embarrassing for all, and could certainly ruin any delicate astronomical observation experiments aboard.

Although one can rely on the inexhaustible imagination of the frustrated human psyche to side step any little space station design flaw, the physics of sexual intercourse in space may be the most discouraging factor of all. It takes only a gentle push with one's finger to propel oneself across the fifteen-foot breadth of the shuttle middeck, and only a little push with the toe to travel up a ladder. If one were floating in the middle of the room, away from the walls, it would be difficult to get anywhere since, unlike in water, there is nothing to push against to get one where one is going.

Given these difficulties, sex in space could take on the comic proportions of a Woody Allen movie, if the lovers aren't careful. To start with, forget about pouring the bubbly to set the right mood. Bubbly doesn't pour in space, and would bring on a phenomenal case of flatulence even if it did. One will have to be content with gazing on the form of the beloved and, as a nation of body worshippers, that may come as a shock. Legs get thinner and chests expand, which is good news for both sexes. This happens as a result of fluids rising upward. However, these same fluids also bloat the face, causing the beloved's features to fill out like a blowfish. But if one has been in space a long time, everybody looks like a blowfish so it'll become the predominant fashion. The gaunt look will finally and happily be out.

Then there is the problem of clothing, the traditional barriers to all healthy passion. Don Juan himself would fumble like an adolescent with buttons, hooks, and zippers since fine motor gestures degrade significantly in space. However, since Velcro has already proven to be the savior of fumblers and little children who haven't learned to tie their shoes, there might be hope of dramatically ripping one's clothing from one's beloved, if Velcro replaces all those little obsolete Earthcouturiered nuisances. But nobody can fling off his or her clothing as people do in the movies, unless one wants to meet up with them again. Clothes will float around and take on a life of their own; they won't lie politely on the ground. The scene at the end of *Moonraker*, in which James Bond is embracing his lovely companion with a blanket chastely draped over them, is totally hokey. More likely, the blanket would be strangling them or hitting them in the face. Also, Bond was a lover who relied on full use of both hands. In space, love will favor one-handed fondlers, those with extreme dexterity in one hand, since the other will be used to help anchor and position the body.

Interplanetary lovers have to keep very, very still or crash themselves through the nearest exit. If they are in an open place, they may thrash around helplessly like beached flounders until they meet up with a wall they can smash into. They can always belt themselves down or, better yet, get some strategically placed bungee cords to help make movements more natural. If they snuggle up in a sleeping bag built for one, which is secured at both ends, there might be hope, though this is an uncomfortable solution, distracting at best. But then, if everything were more earthlike, O space, where is thy victory?

All this depends, of course, on whether they can uncurl their bodies—and keep them uncurled—from the normal weightless posture called the "space slouch." The body tends to curl into a slightly fetal position, which is nice since it eliminates the need for furniture. Everything is standing room in space—working, dining, watching TV—everything. People are quite comfortable in the space slouch. But if one were to try to stand gunbarrel straight in space, it would take exertion, and one couldn't maintain it for long. Undoubtedly, the frustrated human psyche will find a way to sidestep this minor problem too, but it may call for additional exertion or, at the very least, imagination.

The rewards may be worth it. Some space sex theoreticians

claim that at least twenty positions in the Kama Sutra can be attained *only* in weightlessness.

If one were to follow the cliché of the cinema and light a cigarette after all was said and done, every smoke alarm and fire suppressant system would go off. Does the loving couple really want sirens and alarms to go off, and to be sprayed by a fire extinguisher? Then there is the matter of how perspiration sticks in globules to the skin, and makes one feel exceedingly tacky and slippery and, as mentioned before, how baths are rare occurrences. Dressing quickly to avoid detection by a jealous spouse is difficult to do too, unless one can brace oneself in a corner.

Nobody is suggesting, though, that troublesome clothing, difficult posture, alien position, altered physical appearance, smoke alarms, and irate astronomers will stand in the way of the most powerful urge of the species. Experimental mice have already accomplished successful copulation in space, so why shouldn't people?

So far, though, to the best of anybody's knowledge (prurient-minded journalists notwithstanding), mixing men and women in space has been handled very chastely. For one thing, the women who have been sent up are formidably professional, their spaceflights have been very short, and none has been sent up with her husband. The Russians very chivalrously allowed Svetlana Savitskaya exclusive use of the Soyuz for her privacy, fixing up, and so on. The Americans have not been so magnanimous; women astronauts have to park their sleeping bags wherever they can find space, along with the men.

Sleeping arrangements could go the segregationist direction, especially if there is more than one habitation module, or coed.

The thing that almost everybody wants to avoid is a battle of the sexes, especially in an isolated place where there is no privacy. It is always tense and embarrassing to be around couples that fight or even just snipe at one another. Said psychologist Albert Harrison, "Such fighting in an isolated, closed environment would have devastating consequences to the entire crew." Even if they don't fight, there could be problems. Habitability expert Marianne Rudisill of Lockheed pointed out that "Married couples might threaten the welfare of the group because they see themselves first as a pair, a unit apart from the rest of the group."

But the fears may be groundless. Most people tend to take every measure to avoid hostility and conflict in a closed and isolated environment. And it's likely that married couples and families will remain in space for any length of time only after the establishment of a large space settlement with many people. In that social milieu, a family squabble would certainly not have the same dire consequences as in a small group.

Spacecraft habitability experts have more to do than worry about food systems, toilets, showers, and bedrooms. In space, like everywhere else, convenient and pleasant interior design is conducive to a sense of well-being. Houston architects Guillermo Trotti and Larry Bell designed a third-generation space station called "Spacehab" with people in mind. In Spacehab the unpleasant exercise and shower ordeals of space were converted almost to a health club routine in which people could exercise in a wonderfully equipped gym, then dump their wet clothes in the laundry and go through a car-wash shower (sauna, shower, blow-dry, and sunlamp), and pick up fresh uniforms before going to breakfast, work, supper, or bed.

Although space habitats might not have room for such amenities in the near future, space station designers have to make life as convenient as possible in order to not make the environment an additional stressor on an already over-stressed crew.

It's the little things in life that really get to people, the Skylab designers found out. Things tended to migrate into the corners of square pockets, and there were never enough pockets in the Skylab garments. Overpacked drawers are still a tremendous problem in the shuttle, since there is the "jack-in-the-box" effect whenever you open one. Worse yet, it's hard to repack everything again. Housekeeping chores are sometimes made unnecessarily difficult when replaceable filters and uptake vents are inconveniently located. Fans and motors make unnecessary racket, and spacecraft cabins sometimes become too hot or too cold, according to orientation to the Sun. It might be difficult to get the right information quickly on a CRT, and to find the one bit of important information when the proper display does appear.

Noise, temperature, clothing, housekeeping, accessibility to information are all important concerns since failures in these areas quickly raise tempers. For instance, there is a great deal of noise on the shuttle. The astronauts have to shout to be heard across the fifteen-foot middeck, and have to communicate by intercom with the flight deck upstairs. Some had difficulty sleeping at first. The Russians too have a noisy station, but they adjusted to it; in fact, some cosmonauts remarked that they would awaken if even one of the numerous fans shut off at night.

People are very sensitive to temperature. If it is too hot or too cold they have trouble sleeping. The first crews of the space shuttle found the craft uncomfortably cold, especially at night. Now some can be seen wearing shorts. Since blood circulation in the legs tends to slow, long-term spacefarers frequently wear thick socks or heavy fur boots, even to bed.

While the space station designers are laying out the laboratories, workshops, habitation and command modules, they are looking at such questions as: How much noise is generated by the fans cooling the electrical equipment? Should people sleep in little cubicles with their own thermal control? Are the air-revitalization cannisters easy to reach and replace? Isn't it better to create plexiglass drawers with lots of compartments to avoid the jack-in-the-box effect? Shouldn't we have CRT or LCD screens that display information that's color coded—red for essential, and so on? Shouldn't pockets on clothing be round, and shouldn't crews be allowed to embellish them with their own patches to maintain some sense of individuality? Shouldn't we make walls movable to change, expand, or reduce spaces as needed?

After having to share a toilet, a shower, and a galley with twelve other people in a small place, any normal person is going to need to get away. Probably the most difficult—and most essential—calculation is: How much space does a person need for privacy?

Unfortunately, there's no easy answer. Privacy is as much a matter of culture, sex, duration of stay, familiarity between crewmates, and personal idiosyncrasy as it is of volume or space. According to Mary Connors, Albert Harrison, and Faren Akins, the authors of *Living Aloft*, "Privacy is a means by which people can regulate their relationships with other people."

Research conducted by Irwin Altman, William Hawthorn, Dalamas Taylor, and Seward Smith indicates that women are more cooperative than men under crowded conditions, but not for long periods of time. It's possible that the more people there are, the less personal space one requires to maintain privacy, whereas having fewer people requires more physical space between them. Friends instinctively respect one another's privacy, whereas strangers typically become quickly acquainted and then withdraw from overfamiliarization.

Disclosure of information about oneself is a means of maintaining social, if not physical, distance. Crowded cultures, like the Eskimos, who must live at close quarters in a hostile environment for months at a time, create a series of social masks behind which they hide. All cultures develop rules of etiquette governing social distance and personal space. Americans and Japanese are likely to feel their personal space violated when Europeans or Middle Easterners stand a little too close to them while conversing.

Crowding is a very individual perception. What appears

crowded to one person may seem cozy to another. Furthermore, spatial perceptions change with time. The Salyut seemed narrow at first to Lyakhov, "but as our floating in weightlessness was becoming the normal state, compartments of the station seemed to us more convenient and spacious." But cosmonaut and space station designer Konstantin Feoktistov told a Russian correspondent that it was the opposite for most other cosmonauts: "When the cosmonauts pass from the transport ship into the station, they immediately are seized by the sensation of ampleness. . . . But everything is comparative. After many months of flight, this was noted by both crews of the long-term Soviet expeditions on the Salyut-6: the dimensions of the station no longer seemed so ample. During the flight one sometimes wants to be by himself."

People often personalize small spaces with mementos and photos, like naval officers do their lockers, as a means of preserving privacy. In space, many of these methods have already been resorted to. Aleksandrov, Lebedev, and other cosmonauts hung pictures of their families and kept souvenirs from their children close to their sleeping bags. Ryumin too noticed that the Salyut appeared like the home of a stranger, until he and Lyakhov decorated it according to their own tastes: "We moved into a 'house' which had been lived in before, and someone before us had arranged things and hung things to suit his own taste. Thus, we decided we wanted to hang our own 'wallpaper.'"

In essence, with the few resources at their command, they personalized their space, made the impersonal Salyut their home with a slight rearrangement of items and the display of memorabilia. They made it home.

Spacecraft designs in both space programs have considered creating true physical private spaces for space travelers. Feoktistov said the cosmonauts have expressed a desire to be alone at times during prolonged flights, and that it is necessary to find a technical solution for this for longer future flights. Strangely enough, although a privacy curtain was requested on an early Salyut flight, it is almost considered a breaking of trust to shut oneself off from one's comrade. When he was questioned about the possibility of future cabins for space crews, Lyakhov responded: "I'm sure that going off during the flight, when there are only two men aloft, and dealing with something which means leaving your comrade alone, would be wrong. Each of you has to be visible and not hid away somewhere. That is like saying, 'Hey, look, I'm here!' In other words we have to see each other all the time and that helps us work together. If you go off and hide somewhere to get on with a job, then, you understand, there is a sort of lack of confidence that arises."

"In space," said Connors, Harrison, and Akins, "crewmembers must depend on each other for their day to day survival. Individuals in a threatening situation prefer to be in the company of others, not only for the acceptance, verification, and comforts that individuals provide each other, but also for safety. As long as a high level of interdependence exists in space, withdrawal behavior in spaceflight is likely to be either curtailed or highly specialized."

Even Svetlana Savitskaya, who had the entire Soyuz at her exclusive disposal, did not spend much time in it. At bedtime, she chose to hook her sleeping bag in the Salyut and be with the men rather than withdraw for the night into the Soyuz, which she could have done with impunity if she had chosen. In essence, she chose to stay in the open common space with her crewmates rather than to withdraw into her own physical private space.

Skylab was palatial compared to Salyut, and crowding was not perceived by any of the astronauts. There was an enormous open area where storage lockers were located and where the rocket-powered backpack now used on spacewalks in the shuttle program was first tested. In their free time, the astronauts loved to do gymnastics there, and ran along the tanks like mice use an exercise wheel in a cage. It is unlikely that such a large open area will exist on future stations, however, since there isn't room to spare.

While the open area offered space to exercise, and was a

"reprieve" from a sense of confinement, the Skylab astronauts would sometimes become a little disoriented in it and preferred the lower decks with their smaller but more familiar spaces. Orientation in space takes time, and cosmonauts must adjust their "local vertical," their sense of up and down, over some days and weeks. Interior cues—colors, architectural features, orientation of equipment—can help or hinder this. Some genius in the Russian program tried to establish local vertical in the Salyut cabin by painting the lower half a dark color and the upper half a lighter one, imposing a clear up and down visual cue; however, all this local vertical was contradicted by placing the sleeping area on what would be the ceiling.

The Skylab astronauts agreed that interior cues were not absolutely necessary. They adapted, even though they experienced some disorienting moments. "I remember going into compartments upside down, but I reoriented myself according to the room's reference points. I never got lost but saw lots of things different ways," said Gerald Carr, "but that was part of the fun of zero gravity too."

Looking out the window is a restful pastime and a means of mental escape from the environment. Astronauts look out windows like people on Earth watch television. What can be more freedom-giving than the sense of soaring over oceans and continents? Every habitability expert has called for lots of windows in future spacecraft since it is one of the few ways an environmental factor can give long-term space workers a sense of escape and relief.

Undoubtedly, interior spaces will be shifted around as we learn more about how groups relate to the environment during long spaceflights. It has even been suggested that the interior walls of the American space station be made flexible and movable, so people can alter the layout for diversion or according to changing needs. It is certain, however, that the exterior will constrain the "floor space," or rather the volume, for a very long time to come.

The prairie had its sod house, and the Antarctic had its Quonset hut. Space will have the habitation module. Such forerunners were easy to construct from available materials, and served the purposes of the inhabitants adequately although not luxuriously.

The same types of constraints govern space habitats. They cannot be built as we build houses and skyscrapers on Earth: by piecing, riveting, or welding sections together. It's difficult to build a pressure hull in space in this way, and there will always be a problem with creating and maintaining seals. So they will be launched as integral units. Thus, the dimensions of the space-shuttle payload bay will constrain the volume and geometry of space structures (suggestions to use the shuttle's giant external tank for a space habitat collapse under the fact that it is much too big to stay in a safe, stable orbit).

Other geometric shapes are possible, of course. The Spacehab architects thought of greatly expanding the "floor space" by using collapsible structures. They could be reefed like a parachute to fit into the shuttle cargo bay, and then pressurized and inflated in space. If anyone has the courage to live in what would essentially be a balloon in space, inflatable structures could be an efficient way to expand living volume. To create sufficient volume for a large population, one would have to think about inflatable structures. Cylinders cannot be strung together indefinitely, or the structure becomes too labyrinthine, and uses up a costly number of launches for the same volume.

Every architect knows that good interior design can offset psychological stresses, so every effort will be made to get rid of the obvious negative factors. Less obvious is the exact layout or series of layouts (or even flexible layouts, using movable walls) that will be best for spacefarers. Whether the sexes are strictly segregated in their separate modules or whether they are mixed is a spatial expression of a social situation. Tiny personal cubicles (or even the virtual absence of a cubicle in favor of a sleeping bag) offset by a large public area where people could meet for discussion or entertainment is another. Space habitats will be modified and enlarged, and, in the course of the years, will come to express the necessary spatial relationships of the spacefaring inhabitants, rather than the theoretical volumetric predictions of earthbound designers.

chapter thirteen

TINKERING

A human being should be able to change a diaper, plan an invasion, butcher a hog, conn a ship, design a building, write a sonnet, balance accounts, build a wall, set a bone, comfort the dying, take orders, give orders, cooperate, act alone, solve equations, analyze a new problem, pitch manure, program a computer, cook a tasty meal, fight efficiently, die gallantly. Specialization is for insects.—*Robert Heinlein, "The Notebooks of Lazarus Long"*

In full view of the controllers sixty million miles away, the debacle unfolded. Each of the Soviet Venus robot landers carried dual television cameras, other apparatus, and a long, hinged arm that was to drive a spike into the Venusian dirt. Accelerometers on the head of the spike would provide valuable geological measurements of the consistency of the soil.

First the cameras were switched on, and their protective covers were jettisoned. These small metal disks sank to the ground within the camera's field of view. Then the arm, stowed in the straight-up position, was released. Down, down swung the instrumented head, carrying the years of hopes of a large team of scientists and engineers.

The camera's lens cover could have fallen anywhere in a large area, and should, by the law of averages, have fallen safely clear of the descending spike. But as luck or fate would have it, the "nearly impossible" happened. The spike descended on its swing arm right onto the discarded lens cover.

After a voyage of many months and many tens of millions of miles, the simple-minded Venusian robot succeeded in determining the hardness of earthborn titanium.

Astrophysicists can determine, by measuring the spectral lines of stars in distant galaxies, that the basic laws of nuclear physics extend throughout the universe. Astronauts and space engineers (such as those who watched the Soviet Venus probe's embarrassing goof) have been able to verify a similar phenomenon on a smaller scale: The laws of human experience on Earth also extend beyond Earth, into orbit. And one of the most influential of such laws is Murphy's.

"Whatever can go wrong, will go wrong," said an engineer named Murphy, several decades ago. Generations of engineers have extended the scope and subtlety of this law, but not its essence. Breakdowns happen, and repairs become necessary. That can be difficult enough on Earth, but it's often impossible in space.

Some of the most dramatic moments of the space age have involved jury-rigged efforts by spacewalking astronauts and cosmonauts to fix malfunctioning space equipment. In 1973, the Skylab-2 astronauts salvaged their faltering space station by manually unjamming a solar-power panel; later they repaired other equipment, in one case by the simple expedient of hitting the side of the box with a hammer. In 1979, Soviet cosmonauts were faced with a similar challenge when a large radio antenna jammed over one of their station's docking ports: despite being within only a few days of ending their six-month space marathon, the men went outside and cut the snagged wires.

In 1984, two major demonstrations of space repair occurred. American astronauts chased down and repaired an unmanned solar observatory, while Russian cosmonauts performed five spacewalks to repair a plumbing break in their space station's propulsion module. Both events marked new heights in ingenuity and endurance, and saved space equipment valued at hundreds of millions of dollars and billions of rubles.

These dramatic spacewalk repairs have been matched with less heralded successes inside space vehicles. Everything from broken wires to leaky coolant loops have been serviced and fixed. In most cases, the equipment had not been designed to allow inflight maintenance.

For long-term space stations of the future, this experience carries several lessons. First, equipment must be designed from the start to be serviced, changed, and if necessary repaired in flight. Second, since such repairs will inevitably (although unpredictably) be needed, the spacefarers must be provided with a full set of repair equipment, from tools to a good workbench to good diagnostic apparatus to the spacewalking capabilities that matured in 1984.

The world's first manned space station was Salyut-1 in 1971, but few details of its maiden twenty-three-day visit have been released. Following the crew's death on the way home, the memory of the mission became a holy legend of Soviet cosmonautics, not to be sullied by any hint of breakdown or trouble. But among the stories that spread into the West, allegedly based on high-level Soviet "leaks" and accounts from radio eavesdroppers, was that a serious electrical fire on the twentieth day or so of the mission had been the final straw in calling the cosmonauts back to Earth—and to their deaths—prematurely.

Certainly the Skylab mission of 1973–4 had more than its share of space-repair drama. Because of a structural failure during launch, the unmanned station arrived in orbit already badly damaged. One of two main power wings had been torn away, and the other was jammed uselessly to its side (the jamming could have been a blessing in disguise, since otherwise it too probably would have been torn away during ascent through the atmosphere). Thermal insulation had been ripped off the station's side, and the temperatures inside reached levels where food, film, and medicine might spoil—and where materials might start giving off toxic gases.

Within ten days, the spacebound crew had been equipped with two makeshift thermal shrouds, along with plans to rehabilitate the crippled station. They proceeded to do exactly that, using jury-rigged tools (including a bone saw and snippers mounted on a pole, to cut the jamming metal strap) in an emergency spacewalk.

A few years earlier, astronaut Walt Cunningham—then assigned to the Skylab preparations—had lost a design argument with NASA engineers. He had wanted the solar panels to be designed to be unfolded manually by astronauts on preplanned spacewalks, and to be stowed securely up until then. The engineers had vetoed the suggestion as crazy, since it depended on risky, unproven manned spacewalks, and since mechanically nothing could go wrong with an automatic system. Skylab experience vindicated Cunningham (but he had already left the program when another pilot was assigned command of the first mission), yet the lesson took a long time to sink in.

Soviet space officials trod the same "learning curve" in their manned space program. Their experience showed striking parallels, although running a few years later.

They, too, were forced to fix equipment that had not been designed for inflight repair. "In the Salyut-6 there arose the necessity of replacing the pumps in the thermal regulation system," chief Soviet space engineer Konstantin Feoktistov explained. "The flight of [Soyuz-T3] was necessary largely because of these pumps. This was critical and difficult work." Feoktistov had been so fascinated by the engineering problem that he had scheduled himself as the mission's third crewmember, until the doctors told him he was, at age fifty-three, too old.

The lessons of the breakdowns were not lost on the follow-on station being developed under Feoktistov's personal supervision. "I assume that it will not be necessary to repeat [such a repair] on Salyut-7," he explained, "but the design of the circuit has still been changed and installation of the pumps is much simpler. There are many such improvements in Salyut-7."

Other experiences had helped drive that lesson home. Aboard the Salyut-6 in mid-1979, two cosmonauts had been preparing to return to Earth after their record-setting six-month voyage. As part of a radio astronomy experiment, they had installed a folded thirty-foot-diameter dish antenna in their aft airlock, where it was opened by remote control. Unfortunately, the wire mesh snagged on a protruding piece of equipment, and when the men tried to jettison the structure it remained hung up. Although they were in no danger themselves (their spacecraft for the return journey was unharmed at the front docking port), the problem threatened to ring down the curtain on the Salyut module. It could not remain under adequate control with extra junk trailing along behind it.

The obvious solution was to go outside and remove the antenna, and in fact the cosmonauts did precisely that. But the action, and the decision to attempt the action, were far more difficult than it might seem. No proper tools were aboard, nor were there handrails or other mobility aids installed on the station's outer surface near the problem area. The men were tired after the long flight, and not extensively trained for spacewalking. Their lives would be placed more in danger by the spacewalk than by running for home.

Yet they, like the Skylab astronauts six years earlier, found ways to make tools, and they found ways to fasten themselves in position to work those tools and cut loose the snagged wires. They found the physical energy and the mental courage, and they saved the space station for three more expeditions over the next two years.

Inside, cosmonauts had also been at work tinkering with balky equipment. By the time Lyakhov and Ryumin were launched in 1979, the station had exceeded its planned lifetime of eighteen months. They brought pliers, screwdrivers, vises, clamps, nuts and bolts, electrical meters, and a soldering gun with tin solder. With this kit, they replaced the head on their videotape unit, replaced cables in their radio system and their bicycle, and did other minor repairs. A later crew replaced control cables in which a short had stalled one of the station's steerable solar panels in a fixed position.

The results were extremely gratifying, since the billion-ruble spacecraft designed for a year and a half in orbit actually lasted five years. "It can be said with complete assurance," crowed chief cosmonaut Vladimir Shatalov with justification, "that the prolonged operation of the Salyut-6 station became possible only due to the cosmonauts."

Such positive experience helped the Soviet space officials later when they were faced with an even bigger crisis in orbit. Late in 1983, a propellant line aboard the new Salyut-7 ruptured, pouring volatile fuel overboard. The cosmonauts closed off enough valves to isolate the failed section, but they were left with a space station on which half the maneuvering rockets were cut off from fuel supplies.

The incident occurred in mid-September, and the men landed safely two months later, leaving the station temporarily empty. The next crew blasted off early in February for another record-breaking marathon (eight months this time). Added atop their already tremendous training program were space-walking drills and equipment familiarization to allow them to repair the damage and restore the station to full operational capacity. The engineers and procedural specialists on the ground had only a few months to develop these tools and techniques.

In April 1984, cosmonauts Leonid Kizim and Vladimir Solovyov made the first of six spacewalks. First they went back to the station's engine compartment and installed work platforms for themselves and their tools. Next they opened a complex of prelaunch feed ports, and installed shunts across several of them. They then installed new piping and control valves, and wired them into the station's control system. By means of a series of pressurizations controlled by a third crewmember inside the station, they precisely located the line break. With this information in hand, ground engineers designed a special pneumatic tool with clenching jaws at the end of a pole, designed to fit exactly to the point needed. In August, a visiting crew brought up the tool and a videotape of ground simulations. On the cosmonauts' last working trip, they used this press to crush the broken section and thus establish a pneumatic seal at both ends of the broken line. By appropriate combination of valve settings they were then able to restore the entire plumbing of the stations's propulsion system. It was practically as good as new.

Compared to this virtuoso performance in rapid space repair, the 1984 space shuttle satellite rescues pale somewhat. The Solar Max repair in April was suspenseful due to the need to snare the satellite in question, but provisions for the main repair work had been made years in advance when the satellite was still being built. And in any case, the crew and engineers had almost a leisurely time period to put the techniques and tools together. The retrieval in November of the errant Palapa and Westar satellites was also a remarkable feat, particularly in the brief time available to develop hardware to grab hold of the objects themselves, but the relevance of such accomplishments to future orbital maintenance and repair is uncertain.

The astonishingly impressive aspect of the 1984 space shuttle rescues came in the area of human flexibility. In both cases—Solar Max in April and the Palapa/Westar retrievals in November—special-purpose equipment had been built to grab onto the target satellites. Neither operation worked, mainly because the target satellites were configured slightly differently than the design drawings provided to the shuttle engineers.

But in an incredible display of virtuosity, the astronauts used their magnificent tools—the robot arm and the jet backpack—and their even more magnificent human ingenuity, flexibility, and strength, to grab and stow the objects.

The lesson for space-station operations is graphic: The human qualities waited in the wings until the more expensive hardware flopped. In the future, expensive hardware can be dispensed with, and the full scope of spacewalker capabilities can be relied on from the start. With backpacks, special-purpose robot arms, eye/brain instrumentation, and bare (well, gloved) hands, spacewalkers can now be expected to accomplish tasks that before 1984 were considered too dangerous and too difficult. The role of people has expanded, and it has made some space operations a lot cheaper!

No longer will spacewalkers have to lash together jury-rigged tools. On the shuttle, there is already an impressive standard toolbox containing tethers, scissors, lights, portable foot restraints, caddies, wrenches, needlenose pliers, diagonal cutters, bolt pullers, vise grips, hammers, probes, lever wrenches, ratchet drives, screwdrivers, winch and rope, block and tackle, pry bars, forceps, tube cutters, tape, straps, and trash bags. Crewmembers (and two are trained for space-walking on every flight, whether it is scheduled or not) have practiced unjamming hinges, pulling doors closed, manually driving locking pins home, and carrying out numerous other repair and assembly tasks.

With space structures too large to be launched already assembled, the coming years will see the need for spacewalkers to conduct routine assembly and deployment operations. Antennas are an example: there are plans afoot for a five-ton geostationary platform with giant dishes, which will be assembled in orbit. "And then because it is precision equipment it will have to be aligned," declared Burton Edelson, formerly senior vice-president with COMSAT and now a top NASA official. "Then it has to be mated to the booster that will take it to geosynchronous orbit." People in space can do this by hand much more easily than an automation can be designed— and funded!—for the same task.

The Soviets also speak boldly of space assembly by space-walkers, and in mid-1984 they flight-tested a manual welding device, supposedly for just such applications. Since their strategy of space station architecture involves modular buildup, their public pronouncements and prognostications are consistent with their apparent needs.

Spacefarers will be spending a lot of their time inside the stations tinkering with balky equipment, performing preventative maintenance, and setting up and taking down specialized apparatus. American space engineers have been proving out aboard the shuttle the tools and techniques that will be needed aboard a permanent space station.

In the large three-ring-bound book called the "In-Flight Maintenance" manual, one of the most important lessons is the use of gray tape. There are rolls and rolls of this tape aboard the shuttle, for use in securing things to walls, ceilings, and even floors. But in a crucial operation such as the change-out of a file-drawer-sized computer unit, the tape has another important use.

Broken space equipment rarely looks broken, except maybe light bulbs, which can be heard to rattle when burnt out. But a computer with an internal short circuit does not flash sparks like dying computers do on "Star Trek." It looks just the same as the new one being readied to put in its place.

So the very first step, say the checklists, is to take out the broken unit and put a big X on the top with two strips of gray tape. Then you can juggle the two units all you want, as long as the one without the X subsequently gets installed.

This may seem humorous, but such errors have occurred on both Soviet and American spacecraft. Once, a cosmonaut removed two air regenerators, got distracted while reaching for the new units, and when he returned to the task he installed the old ones back into the slots. "The oxygen content began to fall," noted a subsequent report, "but it was noticed [by Mission Control] and the error was corrected before any harm was done." Several Soviet embassy staffers in Washington were subsequently observed buying cases of gray tape at local hardware stores—or so the story goes!

The most dramatic and impressive space repair mission to date occurred in June 1985 when two Russian cosmonauts brought back to life a totally dead multi-billion-ruble spacecraft, Salyut-7. After all the effort at fixing the space station's plumbing the previous year, "Murphy's Law" had struck again early in 1985 when during an unmanned flight phase a stuck electrical switch shorted out the station's entire battery pack. Without electricity, the twenty-ton orbital outpost went dead; on board temperatures dropped and both water and fuel tanks and lines froze solid.

Soviet space officials first wrote the station off, but no replacement was available for several years. Then a team of cosmonauts and engineers devised a daring rescue plan. It was complex, risky, and unprecedented, but previous successes on earlier repair missions had developed the team's confidence and capabilities.

So on June 6 a Soyuz spacecraft blasted off with Russia's most experienced cosmonaut in command. The rendezvous had to be carried out with a noncooperative target, using new tools and techniques.

They were then faced with an even greater challenge. Working by flashlight (no power), with oxygen masks (the air had gone foul), in Arctic coats and gloves (no heat), the two Russian spacefarers located the electrical short and jump-started the station. They restored power, then control. They activated the station's life support system and its rendezvous aids, and within two weeks had welcomed an unmanned freighter carrying replacement equipment and fresh food. They saved the entire Soviet manned space program for years to come, and they proved to any remaining doubters that human beings—and human beings alone—are the most powerful force on the space frontier.

Human flexibility, reprogrammability, and dexterity is still only partially appreciated by space engineers, who seem to retain a preference for complex gadgets when a human hand or even a fist will do. As the scope and magnitude of maintenance and repair operations expand, as they are bound to aboard very large space habitats, the design goals must be redirected to exploit the proven—but often unappreciated—capabilities of the spacefarers themselves.

Murphy's Law was formulated by a human being; only other human beings can counteract its effects.

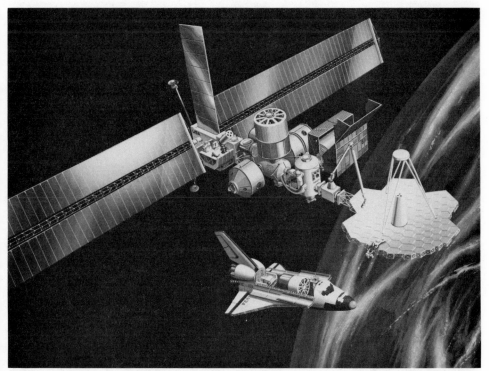

One of hundreds of conceptual designs for a future American space station. Features include solar power panels, instrument platforms, several pressurized modules and docking ports, airlock, "garage" for repair work, articulated manipulator boom, etc. (*Courtesy NASA.*)

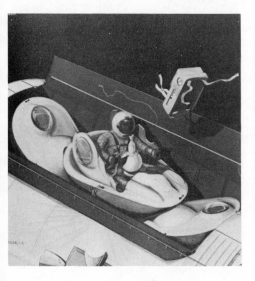

Conceptual design of an individual orbital bailout system shows getting into escape pods and atmospheric entry.
(*Courtesy NASA-JSC.*)

Skylab floor plan was designed around crosswise slices of the big cylinder; smaller cylinders, such as Salyut or Spacelab, are laid out lengthwise. In this mockup, the layout is visible: laboratory at top, wardroom at lower right, bathroom at bottom, three sleeping rooms at left. (*Courtesy NASA-HQ.*)

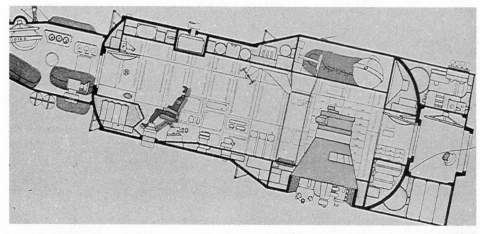

Cross section of Salyut-7 shows overall lengthwise arrangement with local vertical/horizontal layout, except that some interior designer then fastened the sleeping bags to the ceiling.

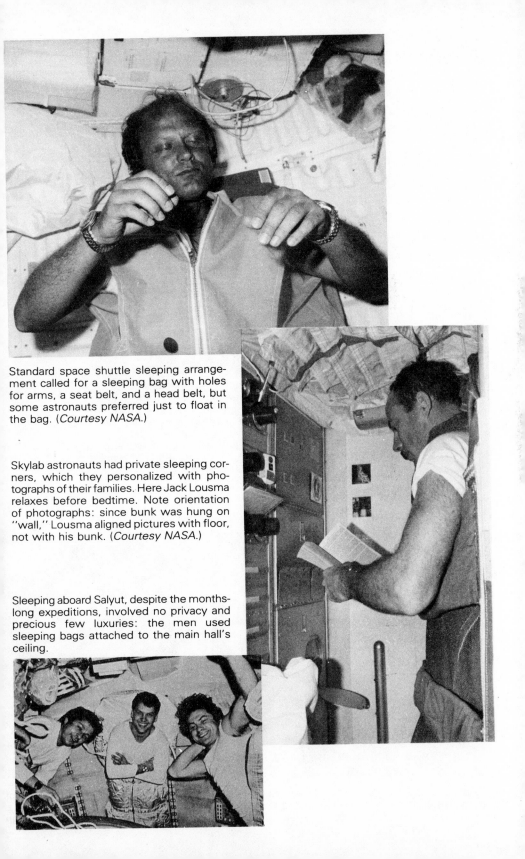

Standard space shuttle sleeping arrangement called for a sleeping bag with holes for arms, a seat belt, and a head belt, but some astronauts preferred just to float in the bag. (*Courtesy NASA.*)

Skylab astronauts had private sleeping corners, which they personalized with photographs of their families. Here Jack Lousma relaxes before bedtime. Note orientation of photographs: since bunk was hung on "wall," Lousma aligned pictures with floor, not with his bunk. (*Courtesy NASA.*)

Sleeping aboard Salyut, despite the months-long expeditions, involved no privacy and precious few luxuries: the men used sleeping bags attached to the main hall's ceiling.

Soviet cosmonauts Valeriy Ryumin (left) and Leonid Popov (right) were thrown together only weeks before their six-month space expedition, yet they were extremely compatible (note body language: similar postures, leaning inwards a bit). "We became as brothers" on the spaceflight, they later reported. Other long-term spacefarers have not always been so lucky. (*Courtesy NOVOSTI.*)

(*Left*) Natalya Ryumina: "When Valeriy told me that he was about to become a member of the cosmonaut team, I cried and tried to dissuade him." (*Right*) Valentina Popova: "Not a single one of the 185 days [of his spaceflight] was an ordinary day in my life. I felt as though I had been set adrift."

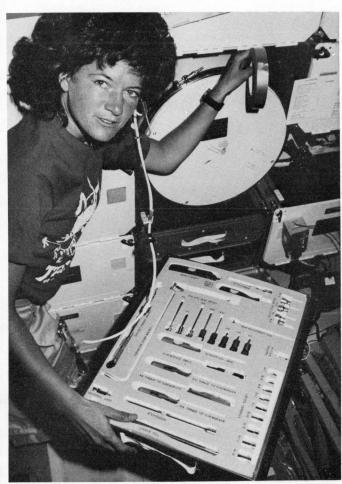

Spacefarer Sally Ride displays one drawer of space shuttle tool kit. Note roll of gray tape in her left hand. (*Courtesy NASA.*)

Cleanliness is next to godliness earthside, but for spacefarers it can be next to impossible. Shower in Skylab (and a very similar one aboard the Soviet Salyut) involved a sealed plastic cylinder which kept water away from rest of station. (*Courtesy NASA.*)

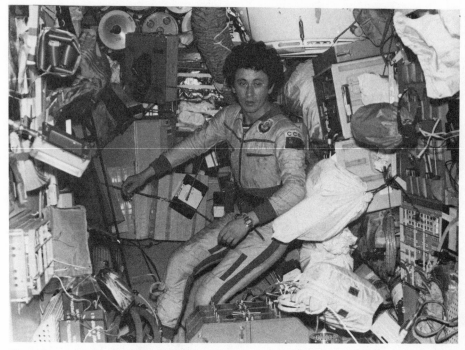

Real picture of a lived-in long-term space station, main hall of Salyut-7 in mid-1984 (with Dr. Oleg Atkov, cardiologist). Every clear surface has been overlaid with gear, charts, tapes, sacks of miscellany. (*Courtesy TASS-NOVOSTI.*)

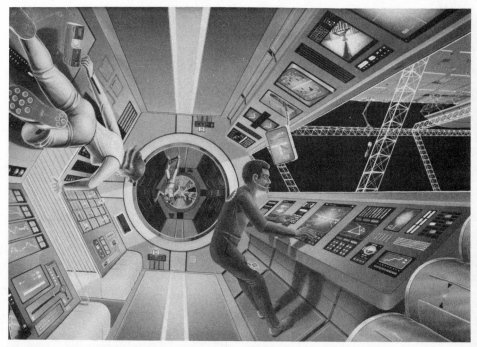

Idealized view of a space station module. Note artwork's no handholds, vast display surfaces, no human touch visible, oversize windows. The painting is obviously a naive tech-nomaniac's sterile fantasy of "people in space" (how grim and forbidding their environment looks compared to Salyut-7's real results!). (*Courtesy NASA.*)

Assembly and repair work in outer space can be conducted by spacewalkers or by robots.
(Courtesy NASA-JSC and NASA-MSFC.)

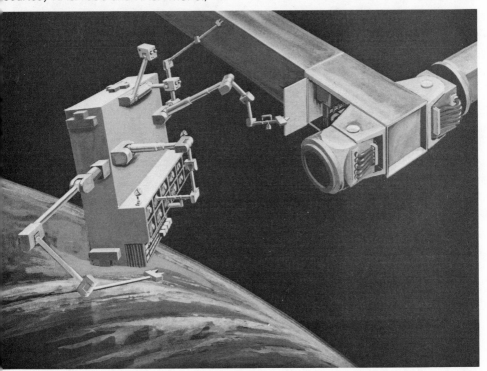

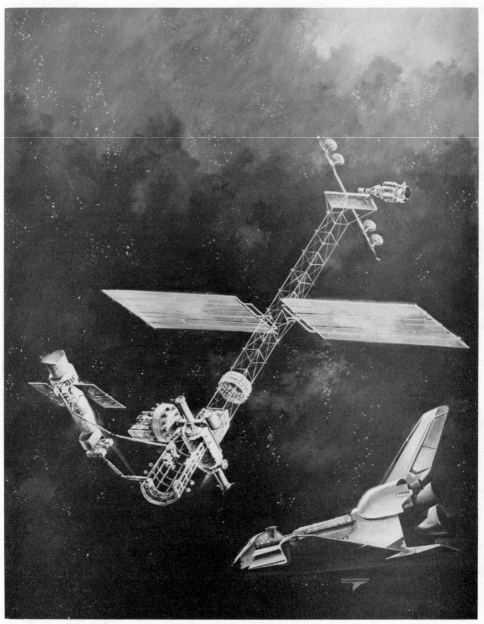

New world for humanity is so alien to Earth that it is bound to engender alien patterns of thinking and feeling. (*Courtesy Grumman.*)

chapter fourteen

ROBOT PARTNERS

People will not be alone when they begin the transformation of space from a void to a cluster of settlements to a civilization. In the past, frontier settlers took along domestic animals to supplement human capabilities for carrying, guarding, and communicating. In space, our domestic animals will be robotic helpers, extensions of human senses, human strength, and even human intelligence.

Scientists now foresee four major incarnations of helpers: expert computer systems and specialized tools; a robot partially endowed with human senses and married to a human operator; semiautonomous robot workers supervised by a human manager; and autonomous robots given some limited human skills, including decision making and judgment.

Aboard current American and Soviet manned spacecraft, automata stand guard like watchdogs over the vehicle's systems. Hundreds of parameters—temperatures, pressures, voltages, pointing angles, switch positions—are repeatedly compared with caution limits. Anything out of the ordinary is grounds for sounding a warning or taking a preset action.

Often, some calculations must be made. For example, a "smart" watchdog should determine if the pressure drop in a fuel tank is due to a slow leak or merely to the tank's cooling down as it faces away from the Sun. Is the trend in some parameter—say, battery charge—such that the value will hit the warning limit after the crew has gone to sleep, suggesting that they be alerted now, before bedtime? These tasks are currently shared by both humans and computers at Mission Control (in Houston or in Kalinin, near Moscow), but future spacecraft must become self-sufficient in such "smarts," simply for reasons of economy and reliability.

A key feature of the man-machine partnership in space is how they interface. How "user friendly" is the automatic system? Does it converse in the soft, genderless tones of HAL, or does the spacefarer have to make his or her wishes known by an excruciating sequence of switch flicking and finger jabbing?

When the spacefarer has a command or a query for the computer system, a wide variety of communications technology can be used. The easiest is a row of discrete, dedicated switches, or an alphanumeric keyboard. Hand controllers (the "joy stick," originally "Joyce stick" after a World War I RAF pilot who built the first one), touch screens (or light pens), and "mice" have been very successful input devices.

If the computer has a message for the crew, is it communicated via audible alarms, a synthetic speaking voice, exclamation marks or dancing figures on a screen, flashing lights, or some other media? Combinations of these and similar systems are already widely used in space.

Expert computer systems, regarded as a form of artificial intelligence, are under development to a limited degree for certain specialized space tasks. Expert systems are, by definition, highly evolved computer programs with a knowledge base built on the analysis, knowledge, inferences, and judgment of human experts. The expert system goes beyond "smart watchdog" systems because it consists of both facts—a body of knowledge universally agreed on—and "heuristics," defined by expert systems pioneer E. A. Feigenbaum as "mostly private, little-discussed rules of good judgment (rules of plausible reasoning, rules of good guessing) that characterize expert-level decision making in the field."

One expert system called KNOBS might help plan and schedule activities aboard the space shuttle. There are many conflicting demands made on any flight, but especially on those involving many scientific experiments such as the Spacelab missions. KNOBS would check constraints such as spacecraft position, fuel, position of the Sun, power, and so on to generate typical solutions to planning problems. It could also check human input for inconsistencies and oversights.

Another expert system called "Diviser" can generate spacecraft command sequences for space probes. In the past, it took people months to figure out precisely what commands to give in order to get an automated probe to carry out a certain operation. However, Diviser would look at all the factors—optimum trajectory in terms of fuel and power consumption, safety to the probe, optimum photo opportunity, amount of time—before generating the best sequence of commands that would resolve the various conflicts and make optimum use of the probe. The "expert" part of the system is that it involves a certain amount of judgment and decision making to do these complicated operations. Under ideal circumstances, the human on the ground need only inform the system what he or she wants, and the program generates the planning sequence.

A possible demonstration of Diviser's capabilities might occur during the Voyager fly-by of the planet Uranus or on the Galileo mission to Jupiter, where it might act as a backup to aid human planners.

Data-analysis expert systems built into planetary probes and remote sensing satellites could save time and money. For example, a good expert system in Earth remote sensing could reduce the amount of data to a manageable portion, and even compare incoming data with stored data, pointing up interesting features or changes. This capability would be a tremendous aid to Earth resource scientists, who are virtually buried beneath an avalanche of data—much of it useless—that takes time and money to sift through for the significant material.

Expert systems on space probes could include repair tasks, if a hierarchy of redundancy is sufficiently built into the probe's equipment. If a problem were to come up, an expert system could analyze faults or problems, and make repairs by engaging a backup system, charting operations around the problem (especially if the problem involves some electrical wiring), and so on. The life of the probe might be greatly extended in this way.

On space stations, we would expect machines to carry out specific custodial duties. Such systems would oversee entire spacecraft systems, say, propulsion systems, and predict and pinpoint any upcoming failures. In this respect, Arthur Clarke's computer HAL in 2001 was very realistic. Its primary function was to oversee all the major spacecraft systems, predict failure of components in terms of probability, recommend a course of action, and express itself in human language. However, progress being what it is in artificial intelligence, it's unlikely we'll have a computer smart enough to be a homicidal maniac by the year 2001.

One must be cautious, however, not to oversell the value of expert systems. "To generate an expert system, you have to ask experts 'How do you do that?' and most of them don't know," said Georg von Tiesenhausen of NASA's Marshall Space Flight Center in Alabama. "Most likely, they instantaneously integrate millions of little experiences to come to their conclusions, and cannot explain them. On the other hand, some experts are reluctant to tell just how they arrive at their conclusions too." The knowledge base in an expert system is often expressed in a series of "if-then" rules to form a line of reasoning. However, according to William Gevarter, a prominent figure in the artificial intelligence field who has conducted many NASA studies: "Not all knowledge can be readily structured in the form of empirical associations. Empirical associations tend to hide causal relations . . . and are also inappropriate for highlighting structure and function. Thus, the newer expert systems are adding deep knowledge having to do with causality and structure. There is also a trend toward non-rule-based systems."

Other programs in the shuttle's computers deal with computations that our brains make unconsciously. To move the robot arm, for example, an astronaut operator merely has to direct the linear motion of the arm's "hand" (the "end effector") in its own coordinate system: forward/back, right/left, up/down. To accomplish these commands the computer must determine just how much and how fast to drive the motors at all the arm's joints (shoulder, elbow, wrist). This requires significant amounts of "smarts."

Some problems remain too complex. The specifications for "collision avoidance" require that the computer have a three-dimensional concept of everything in the shuttle's payload bay, along with a concept of the positions and motions of the arm segments themselves, and be able to predict and warn about impending contact. This is too difficult a problem for the current generation of shuttle computers, so human operators perform the task with cautious motion and cultivated foresight. Human intelligence is often more practical than machine intelligence.

The boundaries of this limited human/robot partnership are already being expanded. The second incarnation of computer knowledge and robotics is a human surrogate: a tele-operated robot made, somewhat clumsily, in man's image. It has no legs, three arms of varying lengths, and cameras for eyes. It has complex computer programs to coordinate its movements, but it has no brain. In order for the mission to be successful, the robot's information system must marry itself to the human sensory system, its motions must be guided by human will, its communication with the human mind must be near instantaneous. In other words, the remote-controlled robot must become an extension of the human body like a person's own arms and hands, servants of the mind.

The concept of wedding robot and human in this way is called "telepresence." In the NASA-sponsored ARAMIS (Space

Applications of Automation, Robotics and Machine Intelligence Systems) study, MIT researchers David Akin, Marvin Minsky, Eric Thiel, and Clifford Kurtzman envisioned using such a man/machine hybrid creature for space work.

Because telepresence requires that the person feel as if he or she were at the worksite instead of far away from it, the visual cues would be the most important. The ARAMIS researchers recommended using a helmet complete with color stereo-optic vision because that is the way humans orient images and recognize components. The cameras on the robot would be "slaved" to move in response to the operator's head position, and could provide both panoramic and closeup views of the worksite.

According to von Tiesenhausen, the ARAMIS contract monitor, the effectiveness of such a helmet has already been demonstrated: "In the summer of '83, we drove a teleoperated dune buggy across some sand dunes. The driver had a helmet-mounted display and responded exactly as if he were actually driving the vehicle—no better and no worse."

In space, however, color television cameras use a great deal of communications bandwidth because they display clear resolution across the entire field of vision. This is probably unnecessary since human vision is clearly focused only in the middle of the visual field. "The best would be to have a wide angle camera with maximum resolution wherever you look directly," said von Tiesenhausen.

Graphic, vector, and holographic displays for three-dimensional positioning of objects might also be useful. "Intelligent" vision might be added later so that the displays would be computer enhanced, highlighting essential but somewhat hidden details, or suggesting the best maneuvers the manipulators might perform.

One of the most difficult challenges lies in making the robot move its arms and grapplers as fluidly as a person moves his or her arms and fingers. Reproducing human dexterity in a machine is almost impossible because the human hand has subtle abilities of precision, and immensely complex information networks that can sense force, position, and texture simultaneously. A robot hand with similar capacity would require a motor in every joint, a frenzy of cables to carry information, wires that are both flexible and rigid, and a computer intelligence to execute complex tasks that hasn't been invented yet.

For spacework, however, the ARAMIS researchers concluded that such dexterity is not really necessary. For most space tasks involved in simple repair or retrieval of satellites—grasping and positioning an object, cutting, welding, latching, or operating simple mechanical and electrical connections—two or three mechanical arms with only seven degrees of freedom and a large assortment of hand segments or "end effectors" is considered adequate.

Because nobody yet knows exactly which satellites a telepresence robot might be servicing, the ARAMIS researchers decided that a modular "mix and match" nonanthropomorphic arm system might be best. In it, different shoulder-to-elbow and elbow-to-wrist segments could be attached to the correct "hand" in order to get the maximum use. Short arms may not be able to reach some worksites, and long arms lack flexibility for close work; the relative length of the arms would be determined by the tasks.

In one scenario, a three-armed robot could hold onto the satellite with one specialized grappling arm, and work on it with the other two, similar to a monkey anchoring itself to a tree limb with its prehensile tail and peeling a banana with its hands.

To increase flexibility, however, some anthropomorphic features might be necessary. Arms constructed of stiff but flexible wires that flex and extend like tendons or "mechanical muscles" would greatly enhance the arm's positioning ability, and allow for some fine motor motions as well.

Some tasks might even require fingers, although a battle rages among robotics experts on just how many fingers might be desirable. Minsky believes a human hand is best, and he developed a five-digit robotic hand. Akin thinks four fingers might be enough. However, at a robotics conference in 1980, von Tiesenhausen made an experiment: "I tied a double knot with three fingers of one hand," he said. "That's the most we'd need."

Important sensory information would have to be provided for the benefit of the human operator. Force sensors or "strain gauges" would be very useful since it would be important for the human to know the robot's strength. If an object began to slip from the robot's grasp, tiny microswitches on the gripping surface could inform the operator immediately. And when the robot reaches out for an object, laser ranging devices would provide much more accurate information than the operator's own eyes. According to von Tiesenhausen, end effectors with pressure-sensor pads—the first of many devices leading to telepresence—are already being tested at Marshall Space Flight Center.

In order to free the operator's hands from a complex set of control commands, the robot's gross motor movements could be voice controlled. Simple instructions such as "stop," "go forward," "turn left," and "turn right" might be voice commanded.

Because telepresence requires such a close human interface with the machine, communications would have to be instantaneous. Thus it is unlikely that an operator could remain on the ground and command his fellow robot-worker in space. According to von Tiesenhausen, the three- to four-second time lag in communications as a result of going through ground-switching stations can cause very degraded performance. A space station or the shuttle, however, would provide an excellent base for telepresence since communication there through the TDRSS (Tracking and Data Relay Satellite System) is relatively swift.

The ARAMIS study concentrated on applying potential telepresence capability to repairing the Space Telescope, AXAF

(Advanced X-Ray Astronomy Facility), and a few others, all of which have been designed with manned EVA capabilities in mind. "Basically, there are similar problems in these astronomy satellites," Akin said. "Not surprising, we found that you wanted to have the teleoperated robot to be capable of everything an EVA astronaut is capable of because of the tasks, which would provide a fine performance comparison, and because you get complementary technologies—if one can do it, the other can do it."

The teleoperated robot, however, will not really compete with NASA astronauts because it will basically be used, according to Akin, "where you can't or won't send astronauts." He explained: "We've done student projects here for manned system to geosynchronous orbit and when you look at the radiation environment out at GEO [Geosynchronous Earth Orbit], you wind up having to make a submarine out of lead, and it's not the kind of environment in which you'd care to go EVA for very long. The real market for telepresence is satellite servicing at GEO."

Some have envisioned robots constructing space structures, but Akin is skeptical. He cautions that "people are very productive in structural assembly in weightlessness—much more effective than on dry land." An MIT teleoperator system called BAT (Beam Assembly Operator) was tested in the Marshall Space Flight Center spacewalk simulation pool, and was found to be inferior to human performance. Although the operators were not very well trained, and the device itself was still a prototype, the researchers found some interesting results.

"When you take a person and put them in a pressure suit," said Akin, "at first there's a tendency to spend a lot of time and effort to maintain body position relative to the cargo bay or the satellite. After fifteen or twenty hours, there's a distinctive adaptation to weightlessness, and you don't care what your body attitude is—you instinctively align yourself to the object you're handling.

"The problem we found with the BAT, is that the person is in a one-g environment, but the teleoperated device is essentially in a weightless environment. It's very difficult to make the connection between the very rigid reference frame that the operator is in to this continually rotating reference frame of the teleoperated device. It may be we didn't train enough, and there may be a corresponding adaptation after training."

Engineers hope to flight-test a basic telepresence prototype system by 1987, although the exact design, hardware, and procedures still have to be worked out. "The problem is that, in servicing satellites," said von Tiesenhausen, "they should be built to be serviced. That costs money. So industry wants to have the capability demonstrated before they design satellites for servicing. The irony is that we have to demonstrate the capability on satellites that were not designed for it."

With artificial intelligence, a teleoperated robot could take on certain human capabilities, such as helping to diagnose problems and predicting the probability of a breakdown. More advanced robot workers could be equipped with threshold sensors that activate a machine's predictive and problem-solving capability only when a certain threshold is passed, thus saving it from being constantly bombarded by useless information from its environment.

Robotically staffed lunar factories have always been part of the blueprints for lunar development. Robots could gather resources in the surrounding area, bring them into a factory complex where other robots could refine the material and others manufacture and package usable goods, whether such goods are propellant for spacecraft or building materials for habitats or solar collectors.

However, any robotic involvement must be justified in terms of time, efficiency, and cost, things that must be taken into account by visionary planetary developers. A NASA study called THURIS (The Human Role in Space) researched the shifting relationship between human and robot in the future.

The focus of THURIS was to break down space work into a set of forty or so generic activities, including monitoring spacecraft, deploying antennas or solar arrays, removing or replacing equipment, encoding and decoding data, performing biomedical sampling, gathering and replacing tools, and so forth. Then the researchers drew guidelines for the optimal man/machine mix in space work. Based on performance, cost, and risk, future program managers must choose whether humans need to do the work alone, assisted by specialized technological equipment such as spacesuits and rocket-powered backpacks, married to a teleoperated robot, or whether robots will act as autonomous systems or as workers supervised by human executives. The use of robots won't be questioned if they are involved in tasks beyond human sensory or motor capabilities, such as sensing outside visible wavelength bands, working in toxic environments, or at duties that exceed human strength.

The surprising THURIS conclusion is that even very expensive space man-hours (estimates range anywhere from $10,000 to $35,000 an hour) are more cost effective than teleoperated or semi-independent robots unless an operation is carried out more than a thousand times. The reason is simple. These sophisticated robotics involve large initial investments in development and software, and are not cost effective unless amortized over a very large number of uses. Some costs depend on "technological readiness." Simple specialized tools take a year to develop, from concept design to model testing. Another thirty months are needed for operation capabilities. Complex systems may take ten to twenty years to develop. Instead of blithely assuming that robotics will be prominent in all future space operations, the THURIS group concluded that a cost-effectiveness mode has to be developed for every individual task. For instance, if the task is daily cleaning of a surface, it might be cost effective to have a robotic unit do it rather than involve a human each time.

Some industrial processes will be automated early in space

development. Pharmaceutical manufacture and crystal growth are both candidates, with a minimum of human interference. In future manufacturing on the Moon and Mars, some operations will eventually become automated, especially highly repetitive ones—the types of tasks that are already automated and robotically done on Earth.

Beyond caretaking and industrial uses, robots have some limited usefulness in exploration too. Semi-autonomous robots equipped with limited-vision systems and contour recognition could set out for Mars in the next decade to explore its surface prior to a manned sortie. "A Mars sample return would push our entire technological base forward," said Wayne Schober of the Jet Propulsion Lab's artificial intelligence branch.

According to JPL's Carl Ruoff, there were important but not insurmountable challenges involved with building a rover that could negotiate the surface of Mars. It would have to step around rocks, but it would not have to recognize them as "rocks," merely as "objects to avoid." "A cockroach doesn't recognize whether it's crawling over a rock or a dried up human skull or some other rounded object," Ruoff reasoned, "and by the same token a rover robot doesn't have to recognize an object to avoid it or step over it." However, if desirable, a rover could stop at an object it does not recognize and wait for instructions from a human overseer on Earth.

Deep-space probes and rovers endowed with artificial intelligence might go where manned exploration would be unfeasible. A robotic mission at its best would be a mission to Titan outlined in 1980 during an advanced automation conference. The mission would be complicated, involving self-maintenance, data analysis, automated decision making, and self-learning systems that are now only the stuff of which silicon dreams are made. The robot spacecraft would go into an orbit around Titan and then send out atmospheric probes, possibly on tethers. After gathering information about the atmosphere, it could then send out surface probes. The whole time it would be analyzing and transmitting data, building a world model of an unknown world. If there is a surface to

Titan, rovers could be dispatched for further data collection. Hypothesis formulation would be intrinsic to the probe's machine intelligence. As more complete data were integrated into the probe's brain, new hypotheses could emerge.

Beyond that, some advanced thinkers inside and outside NASA foresee a race of self-replicating robots carrying out space missions of mind-boggling proportions. According to participants of an "Advanced Automation for Space Missions" workshop in Santa Clara in 1980, "Mankind has fallen heir to an incredible treasure trove of nonterrestrial energy and material resources. It is likely that replicating machines will provide the only 'level' large enough to explore, and ultimately manipulate and utilize in a responsible fashion, such tremendous quantities of organizable matter."

The group envisioned factories over many sections of the Moon, starting with a "seed" group of robots that would reproduce themselves either from a storehouse of manufactured parts or from parts manufactured from local resources. Then they could mine resources for human use on an industrial scale that would otherwise take centuries of development. Self-replicating manufacturing facilities turning out planetary rovers could explore Mars or any other planetary body with incredible swiftness. Raw materials from all over the Solar System could be refined and flown back to Earth or planets under development like Mars or the Moon by a large population of self-replicating robots: hydrogen from Jupiter and Saturn, carbon from Venus, water from Europa, hydrocarbons from Titan, a huge variety of metals from the asteroids.

A race of debris-catchers could automatically seek out and destroy debris as more of it accumulates in the space environment. Another possibility suggested by the workshop would be for robot systems to intervene in a climatic or meteorological disaster, by smashing an Earth-colliding meteor, by manufacturing a huge sunshade for Earth cooling, or millions of mirrors for selective heating—if the Earth were showing signs of moving either toward a new ice age or a Venusian hothouse.

For true megaengineering projects such as terraforming,

converting the Moon or Mars to an Earth environment, su-
perautomata self-replicating systems could import breathable
oxygen for a Martian atmosphere or painstakingly release it
from local resources. With such a race of robot helpers, trans-
forming Mars might take only a century, rather than a mil-
lennium.

What types of creatures would self-replicating robots be?
Would they pass on only their own blueprints to their offspring, or would they add on their own experience? Would
they perceive a sense of self and kinship with their ancestors
and descendants in a seed colony? Would their intelligence
be of a human kind, or some other?

When the great physicist John von Neumann first wrestled
with the implications of computer intelligence in the later
1940s and early 1950s, and dreamed up the first self-replicating machine, he was confronted with the problem of organized intelligence, with a means of teaching a machine to
reproduce itself. He immediately considered biological models,
and, to this day, artificial intelligence efforts have had a bent
toward using the human brain as a model for machine intelligence.

The trouble is that we are incredibly complex both in the
way we think and the way we reproduce. Scientists working
on robotic vision must grapple with the fundamentals of human perception. Some vision systems get the robot to scan
an object from bottom up while others rely on top down or
side to side. Some try to get the system to recognize edges
while other try to get it to recognize the position and orientation of blobs of light. One system is trained by being shown
objects and shapes on a TV camera, like flashcards, for storage
and later identification, although different lighting and orientation might render the object unrecognizable in another
context. Other systems try to get the computer to analyze the
object by approximating the surfaces in three dimensions
using triangulation based on height, shape, and planes of light
that intersect. The robot's recognition of the object can result

from a rigid, preset model of objects or the use of special feature detectors (such as line, texture, and holes). Or it can be informed by more general analytical indicators using generalized geometric models.

Attempts to get computers to understand human speech are highly desirable too, since speech is a very swift medium of conveying information. However, machines have difficulty with even the most basic speech characteristics—recognizing where one word ends and another begins, coping with different speech characteristics of individuals, and so on. They have trouble with sentence structure, the meaning of words, the model of beliefs of the sender, the rules of conversation, and the shared body of general information about the world. Under the best circumstances, machine vocabularies beyond twenty words might be available in the 1990s, and, in all likelihood, the first machine language system to appear will be Japanese, since it has only five hundred syllables compared to ten thousand in English.

Assuming we do create very advanced computers with some advanced artificial intelligence, they will become less and less predictable. "Like with people," said von Tiesenhausen, "advanced computers work heuristically, in other words they can do things with some degree of probability based on incomplete information. We do that too. The outcome is based on probabilities. Very advanced machines can make mistakes, and the more advanced they get, the more human-like, the more mistakes they might make.

"We might have very unexpected actions from self-replicating robots. For instance, if one of these robots meets another and it is programmed to keep itself working, it may dismantle the other robot and take its parts for spare parts—in essence, cannibalize it. Or, on the other hand, it might help it and repair it."

Self-replicating machines could produce exact duplicates of themselves, and could even perceive a sense of kinship, since they would have an accurate description of themselves and could easily identify an identical copy. At the automation conference, the questions of limiting the population of a self-replicating robot community, assuming it was not precontrolled by absence of parts or resources, arose.

Under the most anthropomorphic of circumstances, self-replicating robots might have large blank areas in their memory into which new learning about the environment could be self-programmed. They might pass on a portion of this new learning to their offspring, enhancing the chances of survival. Generations of self-replicating machines, then, need not be exact and somewhat obsolete copies of their ancestors, but could grow and even, to a degree, evolve.

Just what information could be dropped or passed on from generation to generation is as complex a question for a machine as for a human being. Consider the human being. Apart from such basic skills as motion, speech, a generalized world model, certain values, and so on, how much is really passed on from one generation to another? Undoubtedly all our forebears knew how to butcher a hog and preserve its meat, and to milk a cow, although very few of us do now. Lessons are missed and changed from generation to generation, and in machines, the process would be far more rigid since it is a mechanical transference of knowledge.

Whenever very advanced robotics systems are discussed, whenever they are seen not as automata and mechanical partners but as creatures made in our own image, Darwinian terminology invariably crops up. The Santa Clara symposium participants used words like "evolution," "natural selection," and "mutations" when they spoke of the highly developed robots of the far future. And space, like everywhere, was seen as a place with its own unique ecological niches. "Some machines may not thrive as a result of mutations or incurring a hostile environment," said von Tiesenhausen. "On the Moon, if some robots got stuck wandering in a dark crater where their power supply would be diminished or where necessary resources were absent, they might just die."

Nonetheless, all these concepts might be barking up the entirely wrong tree. Machines are not yet, and may never become, much like human beings. There may be a kind of artificial intelligence that is indigenous to machines, and whose organization is completely different than that of human intelligence. For thousands of years, we tried to fly like the birds, by flapping wings, but bird flight was the wrong model. We needed to find another, entirely different, model. And we found, in the laws of aerodynamics, that there are more ways to fly than just like the birds.

A machine intelligence that performs the brain's higher functions completely differently might be sought someday precisely because it does things differently and can overcome human limitations. "Our brain does have limitations, for instance, we can't imagine a four-dimensional cube, and we can only do one thing at a time," said von Tiesenhausen. "But right now our brain is the only model of intelligence we have. One of the cardinal rules now is that every machine must have a human interface. If we built the machine using some other form of intellgience, could we communicate and interface with it?"

Such issues of machine versus human intelligence, development, and capabilities raise fundamental existential questions of human nature, human evolution, basic humanness. In trying to duplicate our own processes of perception and thought, we cannot help but get an appreciation for ourselves, for the little three-pound brain that does so much and still has its fragilities, blank spaces, and weaknesses. Creating future machines in our image, especially for space tasks in which dangerous or remote conditions preclude using human beings, will undoubtedly force us to confront our basic design and methods in order to try to get even primitively functioning machines to carry out our actions. We may truly come to know ourselves for the first time.

For now, though, we're stuck with computers that take twenty minutes to recognize a cup, and robots that do not have enough common sense to know that to fetch a newspaper in the rain, they must first put on their raincoats. "For a long time to come, the human operators bring to bear very important capabilities to space systems," claimed Harry Wolbers, manager of the THURIS study, "especially the cognitive function, memory, evaluation, convergent and divergent thinking, all of which would be very difficult to duplicate in machines. The human being is a very valuable commodity in space, and anything you can do to unburden him—such as using robots in routine, time-consuming, boring activities—is something you must consider."

Wolbers and Steve Hall, the THURIS contract monitor at Marshall Spaceflight Center, agree that in the developmental period of a program, there must always be heavy manned involvement since people are learning about the environment and developing procedures as they go. When operations become better defined and very predictable, machines can step in to unburden the human in the system.

"Robots won't make people obsolete in space," proclaimed Hall. "The relationship between them will be very symbiotic. As our sphere of influence expands beyond low Earth orbit, machines will take up tasks in which humans are apt to make errors. Humans will back up automated systems and be free to take on more complex interesting tasks.

"There might be a series of plateaus in the human-robot relationship in which the humans and robots remain in a state of dynamic equilibrium. This will be caused by the fact that automation technology is developing rapidly and our understanding of human capabilities in space is improving."

By 2010, robots and humans could be working together and apart in many capacities, bringing about an explosion of production and exploration impossible with humans alone. While human assembly workers build large space structures, human factory managers could oversee robotic workers in space factories or direct crews of robot rovers in Moon mining operations. Lone autonomous robots could be assessing the Martian surface in one hemisphere while humans set up their scientific outposts on another.

However, it is an error to assume that robotic development will be linear, in which more complex robots equipped with anthropomorphically organized brains will eventually eradicate the need for human presence. "The question is not humans OR robots," said von Tiesenhausen. "The question is how much participation of each. There's no competition. Every machine has a human interface somewhere. The percentage of participation of each depends on the job. Each task is different."

A century from now, robots of varying complexity, ability, and specialization might inhabit different ecological niches on the space frontier, just as early hominid species with varying physical and mental characteristics inhabited the African savannah millions of years ago. But with a difference: as the new "domestic animals" of the space frontier, they will accompany their creators with directed and purposeful steps, constantly interacting with them and serving the restless activity of their minds and bodies.

chapter fifteen

— The —
MERCHANTS
— of —
SPACE

It is humbling to realize that penicillin was discovered only when Sir Alexander Fleming happened to observe that bacteria did not grow in a portion of a culture that had been *accidentally* contaminated by mold. And synthetic fibers were the result of a DuPont researcher's observation while washing a reaction vessel used in unrelated polymer research. A catalyst critical to the formation of artificial rubber was found when a mercury thermometer being unjudiciously used as a stirring rod accidentally broke. The search for artificial sweeteners was unexpectedly propelled forward when a researcher noticed a sweet taste to a cigarette that had been resting on a laboratory bench. William Roentgen discovered X-rays while investigating cathode ray tubes—and America was discovered when Columbus failed in his search for the Indies.—*Norman Augustine, 1984*

There is nothing more difficult to carry out, nor more doubtful of success, than to initiate a new order of things. For the reformer has enemies in all those who profit by the old order, and only lukewarm defenders in all those who would profit by the new order, this luke-warmness arising . . . partly from the incredulity of mankind, who do not truly believe in anything new until they have had actual experience of it. Thus it arises that on every opportunity for attacking the reformer, his opponents do so with the zeal of partisans, the others only defend him half-heartedly. . . .—*Niccolò Machiavelli,* The Prince, *1537*

Periodically, and for unfathomable reasons, the American government undergoes fits of farsightedness. Every generation or so, a significant portion of the federal budget is diverted into some new scheme of transportation or expansion. Every time it happens, critics bemoan the waste of money in harebrained investments that will never pay off. And practically every time, usually long after the politicians who sponsored the expenditures are gone and forgotten, the spectacular payoff enriches and invigorates the nation.

In the first decade of the nineteenth century, it was the "National Road." Proponents wanted to build a paved highway west from Maryland through the Cumberland Gap into Ohio and beyond. Critics declared it a waste of money, a "road to nowhere," at the other end of which would be found only savages and desolation. The distances were too great, the open spaces too vast, the prospects too terrifying for any reasonable person even to dream of following such a road. But the proponents carried the day, under the slogan of "Let us conquer space." The road was built, and commerce flowed east and west in the decades that followed.

In later eras, Washington allocated funds to canals, riverboat navigation aids, railroads, and other struggling, fledgling transportation schemes. Early in this century, the equivalent of tens of billions of dollars were spent on the Panama Canal, ostensibly for military reasons, but its commercial value soon eclipsed such rationalizations. The postal service underwrote aircraft development in the 1920s with its subsidies for airmail.

Not every investment has panned out: the dirigibles of the 1930s and the nuclear powered merchant ship of the 1960s fell victim less to technology than to a loss of nerve, but they fell from grace nonetheless.

And what do today's critics say about the government investment in space transportation? In an unconscious parody on all the naysayers of the past, they recycle the same arguments, even the very same phrases. The space shuttle, sneered Nicholas von Hoffman, is "a ferryboat to nowhere."

But the cynics and the bottom-liners are, as usual, wrong. There is already an embryonic space economy, currently involved in communications, pharmaceutical manufacture, and transportation to and from low Earth and geosynchronous orbits. For the bottom-liners, the penny pinchers, the uninquisitive, soulless, and humorless Scrooges of the world, this is what the Dow Jones in 2085 will look like.

In order to have a space population, people and equipment have to get there from Earth. Launch services are already an area of key development and competition, and will develop as trade between the two spheres enlarges.

Launching rockets is a complex business requiring long experience and a firm grasp of rocketry and propulsion. Mistakes are costly. When Space Service Incorporated attempted to test the Percheron, the world's first privately built rocket, it exploded—along with a few million dollars invested. "They say you have to blow one up to come to the party," SSI's president David Hannah said philosophically, "and we came to the party."

Potential customers are, of course, the numerous corporations and nations launching communication, Earth resource, and weather satellites, and later, some forms of commercial manufacturing facilities, small factories and the like. Workers, settlers, and tourists must be carried by a large vehicle like the shuttle. Many airlines have already expressed interest in using the shuttle as an airliner, and it's possible that one of them or a consortium will buy, lease, or rent a shuttle for that purpose when the space traffic justifies it.

If we begin to live in low Earth orbit, geosynchronous orbit, or on the Moon, Mars, and the Asteroid Belt, there will be crucial lines of supply among these outposts and Earth. Transportation among them will be vital, and a lucrative space trucking industry would be essential.

Federal Express, UPS, Emery, and Airborne all have a future in space, too. People living in remote space outposts will be able to take along only a limited number of personal items because baggage costs a fortune to launch. But packages from home are important psychological boosters, and families might gladly pay the additional cargo charges to send up special items, especially on a birthday or holiday. Transporting cargo of all types into space, especially in the form of unmanned vessels that can just rendezvous with a station, will be big business.

Where there are highways and travelers there will also be breakdowns and repairs. Towing services for satellites, freighters, propellant tankers, and personnel carriers would surely find their place in the space economy.

Communications was the first big space industry, and is likely to remain so for a long, long time, since satellites and technology are continually being upgraded and enhanced. In space development, communications will play an important part. Earth sensing and automated or robotic operations on the Moon or in geosynchronous orbit require a very high rate of data return. If space becomes a busy place in which several activities requiring communications are occurring simultaneously, communications will have to be expanded to meet the demand.

As population grows in space, fresh demands on communications facilities will be made in the form of business and personal communications. When people live in space a year or more, they may wish to make more calls home than the company or station allots, and the same might be true of mail services beyond a certain size and weight. A premium is already placed on communication with family by the Russian cosmonauts. And considering that they speak to their families only once a week in a somewhat public atmosphere, think of what a luxury it would be to have a private conversation with one's family whenever one felt like it—in a moment of deeper than normal loneliness, or at the end of a trying day. Space station and lunar developers, research scientists, and so on may pay gladly for these additional services. In essence, when they reach out and touch someone on Earth, Ma Bell may be there with the open palm unless some clever communications officer can think of some way to hitchhike additional personal calls on scheduled downlinks of data.

Although it's difficult to think of real estate in space, certain areas of space are already more valuable than others. Geosynchronous orbit is such a place. From there, one can watch one of Earth's hemispheres continuously. Continuous communication can be established between geosynchronous orbit and points on Earth. People can and do aim their fixed dish antennas at satellites in geosynchronous orbit, and never have to invest in expensive tracking equipment. And even though geosynchronous orbit is a huge arc, a limited number of satellites can be located there; if there are too many, their radio signals could interfere with one another.

It's probable that areas in the orbital plane and not too far from a large established space station will be valuable too, since cargo vessels and shuttles from Earth can more easily rendezvous with them than with structures in odd orbits.

Also, structures built in space are, in a sense, real estate. Free-flying unmanned platforms will lease out "space" like an apartment building. The same will be true of some space stations. Hotels will be big business because space offers such a great view. Hotels with good restaurants, gambling, shows, luxury shops may be visited by long-term space workers starved for sensory stimulation and diversion, as well as by Earth tourists.

On planetary surfaces, it's essential to locate mining operations near abundant resources that are easily accessible to incoming and outgoing spacecraft. Valuable areas would also be those where some supplies vital to humans can be reached. On Mars, you'd like to locate bases near resources vital to people, not necessarily vital to mining interests. Antarctic development has done so.

Some people think that valuable minerals might be found on planets, worthy of sending prospectors on an interplanetary gold rush. The asteroids appear to be promising sources of useful minerals such as nickel. However, the real raw material bonanza may be in materials that are abundant on Earth— oxygen, hydrogen, and water.

These materials will not, of course, be imported directly to

Earth, but will remain in space for use in building a space economy and space structures. Massive items like water and building materials are very costly to lift off the surface of the Earth. At $200 million per shuttle launch, it would take numerous launches to build a large space garage in which to service satellites. Propellants for refueling OTVs and resupplying free-fliers and space stations are also very heavy and very costly to transport from Earth. Many engineers, mostly Moon buffs, have pointed out that propellants, construction materials, power stations, and a host of other useful items are more easily mined, processed, and transported from the surface of the Moon. For large space developments, the cost would be several orders of magnitude smaller than importing materials from Earth.

Just how large a space economy needs to get before lunar mining becomes profitable is, of course, debatable. Hu Davis of Eagle Engineering, NASA's Wendell Mendell, and a number of experts who have looked at lunar economics feel that modest-sized, simply designed Moon factories can be marginally profitable fairly early on. Using Moon dirt for insulation around some satellites in geosynchronous orbit, for instance, would probably extend the life of the electronics on these satellites as much as 100 percent. When the space station is operational in the 1990s, and OTVs are used to fetch and repair satellites, lunar-refined propellants become a consideration. Solar power beamed from the Moon back to Earth would always have an attractive side as fossil fuel deposits become more and more depleted. Current NASA projections point to developing lunar resources early in the next century, certainly in plenty of time for all three economic factors to develop a need for those lunar materials.

Key exports to Earth will involve pharmaceuticals, small-yield metallurgy, and industrial crystals. The products have a high price and a small weight, like the spices that opened up trade with the Orient in centuries past.

Other industries, as yet undiscovered, will undoubtedly

surface in the years to come. Luxury items and *objets d'art* manufactured under unique space conditions might prove profitable. Something gimmicky like an ultrapure space-produced perfume might be a candidate. *Objets d'art* would almost certainly include excellent glass and metal sculpture, since these materials come alive with new potentials: perfect spheres, for instance, are possible only in space; new alloys can be melted and set free to form jewelry or sculpture.

The fundamental law of economic predictions is that you can bet that the biggest fortunes will be made on something we don't need now but won't be able to live without once it's available. Witness the phone, electricity, the refrigerator, toasters, television, tobacco, and tea.

Health is big business in the United States, and it's possible that space weight-loss clinics might cash in on the natural tendency to lose weight temporarily in space (though there's a certain amount of quackery involved here too). It's conceivable that a recuperation center could be set up to cater to people in extremely debilitated state, although the individuals would have to be well enough to withstand the g-forces of a launch.

Food for space stations and lunar bases will be catered from Earth for a long time, probably at the expense of the agency or company that is maintaining the space facility. Very likely, private caterers experienced in packaging and advanced preservation techniques would be hired to provide this service, just as they are now for airlines.

However, on space stations, the food tends to get boring, especially the types of foods ideal for space use: unrecognizable "extruded," preserved, or highly refined food. The cosmonauts hankered for the fresh strawberries sent up from Earth, or fresh onions that they grew themselves on the space station. Special foods with flavor, texture, and crunch would find huge favor in sensory-deprived space. Because it is so costly to launch large volumes of these items into orbit, and because their packaging generates a tremendous amount of refuse, an orbiting space garden might be an early profitable industry in space, provided that space stations are not themselves routinely outfitted with space gardens. It is difficult to say exactly what size population is large enough to make such a venture profitable, but certainly a space culture of, say, one or two hundred people could make the industry a going proposition. Again, companies or nations or agencies in need of fresh supplies could buy directly from space farms, or workers could spend their money on such luxuries to supplement treats that their employers do not provide.

This leads to interesting questions: When the first dollar is spent in orbit (they're carried aboard only as novelties now), what will it be spent on? Food? Entertainment? Clothing? Cosmetics? Alcohol? Phone calls? When the first item in space is stolen, what will it be?

chapter sixteen

— The —
SOARING
SOUL

In the course of daily life, people enact a variety of different roles (teacher, husband, father) with those who enact interlocking or recip- rocal roles (student, wife, son). In addition to providing stimulation, such diversity allows the person to exercise different skills and talents, and may be important for a complete sense of identity. Variety in relationships also provides the opportunity to formulate a balanced response, as when a dispute at work is placed in perspective in a session with a spouse or a friend.

Participants on long-duration space missions will be temporarily, or perhaps permanently, extracted from the ongoing relationships that are important to them, and will be thrust into a microsociety which will impose its own social deprivations and hardships.—*Connors, Harrison, and Akins,* Living Aloft, *1985*

Many is the child that journeys this high to be different, planning to get from the mountain something their natures couldn't get them below. That comes to nothing. Can't cheat the mountain, pilgrim. The mountain's got its own ways.—*Will Geer,* Jeremiah Johnson

When *Pravda* printed excerpts from Valentin Lebe- dev's diary of his 211-day flight, they titled it: "His Heart Remains on Earth." Nothing could be more true.

When Lebedev blasted off from Earth, he, like all his com- rades who have spent months in space, was uprooted from all life as he knew it: from spouse, children, parents, friends, co-workers, home, office, friendly neighborhood hangout— even from the days and seasons, the rhythm of life. To people who have always dreamed of floating in space and working among the stars—and who have never been there—this would all seem like "no big deal" (nichevo, in Russian). But it turned out to be a very big deal indeed. Lebedev called his launch from Earth "like breaking a chain"; but it was a chain from which he never could be free.

"All the inexperienced cosmonauts agitate for longer and longer spaceflights, but the veterans of long-term flights don't," said Oleg Gazenko, head of Soviet space medicine. "Ryumin [the current record holder for spending the most time in space than anyone else] hates spaceflight."

The reason is simple. In many respects long-term spaceflight as it exists now is a bit like jail. The stages of psychological adaptation are almost exactly those of people in other confinement situations, on submarines, Antarctic expeditions, and prisons.

According to Connors, Harrison, and Akins, there are three stages of reaction to prolonged isolation: heightened excitement and anxiety; a day-to-day routine that may be characterized by depression and regret; and anticipation of the end.

Cosmonaut Lebedev claimed to have experienced nine. The first stage begins in the weeks and days before launch. This time consists of long, intense training in which astronauts and cosmonauts alike are drilled on every possible emergency contingency, allowing a certain amount of "worry work" to be done.

The tension frequently spills over into their family life. "It's difficult being the wife of a cosmonaut," Natalya Ryumina told a reporter for *Soviet Life*, "because the profession [of cosmonaut] is so exacting and demanding. Not just the flight but each training session, of which there are a great number, is a physical and emotional strain on the cosmonaut. It's not easy to be constantly aware that you have to help him ease the tension and relax."

In private moments, between training sessions and away from his family, a cosmonaut's mind may range across a wide assortment of potential problems. Valentin Lebedev spent a spring evening just before his launch strolling beside the riverbank in Tyuratam, and thinking. "I thought of the work and the long years to my second flight," he recalled, "the labyrinth of human relationships, the difficulties and so forth. And now I had reached the summit, and I was frightened, but not from the height that lay ahead. Not from the dangers that are possible in our profession. Not from the difficulties of a long mission and the great amount of work; but from myself: was I capable of living and working for such a long time with my comrade? Could I manage? It would seem that the main difficulties were behind but they are turning out to be ahead."

Ryumin, too, prior to his first long-term flight had had similar thoughts: "During the rare few moments we had to ourselves before liftoff, my heart would fill with chilling doubts: Would we succeed? . . . A flight is always a risk. . . . We are always aware that we have been entrusted with a tremendous task, to complete the work of large collectives of workers, engineers and scientists. We, the cosmonauts, are the final link in this chain which begins with the conception of an idea by the project engineer and ends with the written account after the landing . . . . And so this tremendous responsibility for the work of the entire collectives, on the one hand, is encouraging, but on the other hand, becomes an awesome burden."

Ryumin had very little time to worry about his second long-term flight since he had been assigned to it only a few weeks before launch, as a result of another cosmonaut's injury. Instead, one April day he considered the implications of leaving the Earth again for a long time: "The weather at the cosmodrome was beautiful. The days were warm and sunny . . . . I felt a slight sadness in that I could not see what lay in front of us. And I love summer. It seemed that if the flight had been scheduled in the winter, it would have been better."

Saying good-bye to spouses and children is hardest of all. "My mood is complex," Lebedev wrote in his diary the night before he left for the launch center. "Vitaliy [his son] caresses me and kisses me; he senses, dear boy, that his father is flying off for a long time . . . . When we drove away from the house, I looked to the balcony and saw mother wiping away the tears. I waved to her; she did not see me."

Aleksandrov's children loaded him down with souvenirs to carry with him into flight. He cherished these deeply, and kept them close to him. "Natasha [his wife] asked the children to get together some presents for their dad, but with the condition that they be small. Serezha found a teddy bear and Irishka sent a small doll. I still have the dried violets which they sent from home before the launch."

Launch days hold their own special terrors for the families of space travelers. For instance, Larisa Yeliseyeva's husband, Aleksey, had three very short flights, the longest only about five days. But she had special problems because Aleksey's flights were always associated with death and disaster. His first flight should have been in 1967 when his crew would have met another ship, spacewalk to it, and later land in it. But troubles developed on the first ship, and Yeliseyev's launch was cancelled. The troubled ship then crashed on landing, killing the pilot, Vladimir Komarov. If Yeliseyev had been on board as planned, he too would have died.

His next flight was marked with a failure to link up with the space station. On his third flight in 1971, Yeliseyev was supposed to spend thirty days aboard Salyut 1, but again, the link-up failed. The crew which followed him—Dobrovolskiy, Volkov, and Patsayev—managed to link up but died on their return to Earth when air leaked out of their Soyuz capsule during their descent to Earth. Just as in the earlier flight, it could have been Yeliseyev's life that was lost. He had stood closer to death than any other cosmonaut since comrades immediately preceding and following him in flight had died. Only the slightest alteration of fate could have put him in their place.

His wife, Larisa, was understating her anxiety when she said: "I took all three flights very hard. The most difficult part is when your husband leaves for the cosmodrome. All the reasoning in the world didn't help me when Aleksey made his flight."

Even for the men, there is the unspoken but never unrecognized thought that the good-bye might be forever. "In an unproven and untested technology, you never know what kind of failure awaits you in weightlessness," Ryumin wrote in his memoirs. He confided to veteran journalist Nikolai Zhelevnov that, prior to flight, one "pays the bills in advance, by psychologically preparing himself. . . ." Immediately prior to launch, the cosmonauts must put aside their fears to concentrate on the work ahead.

During liftoff they may briefly acknowledge the magnitude of their endeavor in some unconscious part of their minds. Ryumin used the words "torn away from Mother Earth" when he described his launch. He told *Soviet Life:* "As for there being another me, I do feel that I am another person when the rocket blasts off from the Earth and loads increase and it seems I am leaving the planet forever and that there is no power that could bring me back."

But there is not much time for reflection since the spacefarers must maintain their concentration and efficiency until they establish a routine. It takes a long time to enter that stage of confinement. They may be ill the first few days. Afterward, it takes time to coordinate their body movements efficiently. The space station has to be rearranged for the mission and new experiments have to be set up. All this might take a week or ten days, longer for some crews.

The focal point of a person's entire life in space is work. Imagine what life would be like if you worked and lived in the same spot with the same person day in and day out for six or seven months. Work's importance is expanded far beyond its normal proportions.

"During a mission the crew carries out the most diverse kinds of work and it has a lot on its plate," said Berezovoy. "And if something suddenly breaks down it is very easy to take the attitude of, well, it is not my fault. Of course, no one is about to blame you, and you are, in a way, out of it all. It is easier to blame circumstances. But the thought is always gnawing at you: when you get back to Earth will you see it the other fellow's way if you have indulged yourself? For some experiments are prepared by an entire collective. You do something not quite right, and years of work are lost. This is why we sometimes worked at full stretch for 14 to 16 hours a day, surpassing ourselves. Discipline is very strict, you cannot weaken. Just fail on one command and all the work done by you and by others could be for nothing. On Earth you often do not see the final result of your own labor; this dependence is covered, hidden. In this case everything depends on your conscience. But in space both your work and your conscience are on view for all, and you cannot hide behind others."

Some of the experiments are interesting to the spacefarers, and some aren't. The most interesting ones are those which allow them a certain amount of creativity. Too often in both the American and Soviet programs astronauts and cosmonauts are mere technicians running other people's experiments with little input of their own. The Russians found that when asked, the cosmonauts were eager to offer their own evaluations. Some scientific groups even awarded them honorary degrees, and gave them a sense of creative participation in their work.

Lebedev pointed to the type of person who would thrive in space: the workaholic. "I would want them [future space workers] to prepare for every experiment, every effort, not as operators, but as engineers—using creativity. . . . In order to feel more at ease during long-term flights, they must find their passion and live it in space."

The pace of the work is hectic. According to Gazenko in an interview with Russian journalist V. Yankulin, "For cosmonauts, free time is rather conditional. . . . It is necessary to take into account the specific nature of the time organization of the cosmonaut's activity on board: a great many operations are imperatively tied in to definite moments in time. They can be carried out only at a particular time. The position of the station in orbit relative to celestial bodies, to definite points on Earth's surface, to land observation and data transmission stations—all this is tied in to a rather rigorous time schedule, and this in itself also exerts a definite pressure on man. A cosmonaut, in essence, sits on the end of an enormous second hand rotating on the dial of a circumterrestrial orbit."

As a result astronauts and cosmonauts frequently use their free time to finish things that were left unfinished, or go over the next day's tasks. In this way, their leisure time acts as a kind of safety valve into which the pressure of uncompleted tasks can spill. "There is no pause for emotions," one of them wrote. "Other work is waiting. This is something like work on a conveyer belt, but with the single difference that on a conveyer belt you stand in one place and perform one and the same operation, whereas here you seemingly move on some time transporter along the conveyer belt and perform different operations. That is why the cosmonauts in many cases use their free time in order to do that which they had not finished or were forced to postpone."

In the meantime their emotional life is on hold. Even though spacefarers get to know each other well, they very rarely become as brothers. "We rarely talked about things other than work," said Ryumin of his 1979 voyage with Lyakhov. Said Lyakhov of his later mission with Aleksandrov: "The main thing is to fulfill the program, and this means work, work, work. Quite honestly, there is little time for emotion." His spaceshipmate interjected: "Or more accurately, there is no time at all."

On Skylab, too, the astronauts got along and worked well, but did not become too close. Both Russians and Americans instinctively avoided one of the pitfalls of confinement: overfamiliarization. By distancing themselves and withholding personal information, they respected one another's privacy.

Internally they tried hard to suppress emotions, to keep themselves in line. Said Berezovoy: "I learned from the fellows who had been on long-term flights before us that certain critical moments arise during orbital flight when one must 'soundly dress himself down,' not allow himself to 'slide' so that work goes on as scheduled."

Two months into the flight with Berezovoy, his crewmate Lebedev wrote: ". . . There are tense moments between the crew, but no outburst can be permitted. Otherwise, if a crack appears, it will get wider."

Two rookie cosmonauts were on the first successful civilian Salyut mission early in 1975. Besides the burden of pioneering a technology that had recently killed a crew, the men were suffering under the "floating days" rhythm, waking half an hour earlier every day to keep pace with the station's orbital shifts (that plan was scrapped a few months later). Mission commander Vladimir Gubarev later described the additional psychological stresses they encountered: "The first three days of our work in orbit, the relations with Georgiy were virtually the same as on the ground, where we were friends. They were businesslike, friendly, just as healthy as they could be, as required for normal work under these conditions. Several more days passed, and we sensed the atmosphere was becoming charged. Sometimes, we differed in assessing the same event. I soon noticed that Georgiy became even less self-controlled, abrupt and wound up. On the ground he was notable for self-control, modesty, and calm. We felt worse and nervous tension increased to the limit; there were some manifestations of discomfort."

Gubarev stressed his surprise: "The most interesting thing is that this had never happened with us on the ground, in the simulator. We both tried to overcome the 'new nervous state.' We had to somehow smooth out the rough edges, forgive, and bear with such deviations in the actions and behavior of our partner. We accomplished much." They completed the thirty-day mission as planned.

Despite these spacefarers' deliberate attempts to smooth things out between them, there usually are no "rap" sessions in which differences are aired openly since the confining space environment enforces a suppression of hostilities. Cosmonauts avoid competition and conflict at all costs. Even the chess set that was brought on board Salyut-6 was left untouched, except to play against the controllers on the ground ("Earth" versus "space"). Nonetheless, long-term spacefarers get to know one another well through a hyper-awareness of one another. Said Ryumin of his flight with Lyahkov: "Up to the time of the flight, we trained for a long time together, but our preparation took place among a circle of people. This is not quite the same as finding yourself as a twosome. Here every word had a meaning, even the tone was important. It was important to really be sensitive to your friend. . . . to know, to understand his mental state and to analyse the possible consequences of your communication, your words."

Their most intense emotions are still reserved for their families on Earth, which they keep to themselves. Lebedev hung a photograph of his little Vitaliy above his bed like an ikon: "Each evening I kiss the photograph of my son. He looks out at me so kindly that when things are difficult I say to him: 'It's nothing, son, I'll manage. Papa will not let you down.'"

When a visiting crew brought up a thick envelope of letters, Lebedev waited until everybody was asleep and then went into the private transfer section of the Salyut to read them. "Lyusya [his wife] and little Vitaliy had written me many letters. Then I lay down in the sleeping area, but I could not close my eyes for thinking about the letters and what we would say when we met."

Aleksandrov never left his family behind emotionally, since some of his emotional rhythms were based on his communications with his family. His memoirs are heavily punctuated with his thoughts and feelings about them:

"July 6—Today is a holiday. I received a letter from Natasha [his wife] today via teletype.

"July 9—We conducted a geophysical experiment and then talked with our families. Serezha lost a tooth. He tried to explain that they had harvested the crop on the balcony: one small cucumber which he and Irishka ate . . . . [They] told family news. Natasha said that all our friends were over to our house. They said many good words and promised to come again after our landing. . . . These encounters, of course, do not go by calmly and the effect remains right up until you sleep.

"July 10—Natasha sent another letter, but it's shorter. The main thing is that they are with us.

"July 15—Today Irishka napped in the nursery and played a little with the children. She is getting accustomed to the collective. This report cheered me up for the whole day."

Meanwhile, back on Earth, the spouses go on with the rhythms of everyday life—their jobs, their chores, their child care worries. But they, too, are sometimes in a state of emotional suspended animation, and feel detached from life around them. Said Valentina Popova: "Not a single one of the 185 days [of her husband Leonid's flight] was an ordinary day in my life. I felt as though I had been set adrift. I'd go out into the street and see people talking and laughing, but I couldn't stop wondering how Leonid was doing out there."

Weekly television visits between the cosmonauts and their families came and went. Russian psychologists noted silently that the sessions with the families got longer and longer as the flight went on; those essential relationships became even more profoundly necessary. The tenuous tie with Earth was more tightly grasped in the busy, sometimes interesting, but emotional barren wastes of space. To use Lebedev's metaphor of the broken chain from Earth, they were trying to reforge some links and find their way back.

Sometimes the cosmonauts try to reach out, fight their sense of helpless isolation, and become active participants in their families' lives. During one communication session Lebedev and Berezovoy staged a birthday party for Berezovoy's eight-year-old daughter: "We made a cake from packets of bread. Instead of candles we used felt-tipped pens, and we simulated the tongues of flame with foil. There were also electric candles: four flash lamps. And so as to make eight, for the number of years, we used the reflection from a mirror. We hung various colored balloons that scooted about on the vacuum cleaner, and a ball." This touching attempt to play clown at a child's birthday party while orbiting two hundred miles overhead succeeded: "In all, we quite cheered the young girl by television," he reported with immense satisfaction.

Milestones in life are passed in space as they are on Earth, and it seems important to mark life's continuity with a little ceremony. Birthdays are carefully celebrated with bits of poetry from co-workers on Earth and saved-up delicacies in space. In a diary entry on his fortieth birthday, Ryumin reviewed his life: How his parents had helped establish and build the town in which he was born, and later moved to Moscow; how he worked as a design engineer in the unmanned Moon-probe program for the father of the U.S.S.R.'s space program, Sergey Korolev; how he helped design the first Salyut. He went into the cosmonaut corps, was selected, and worked years in preparation. He thought about his wife, how they met, how she understood the problems of flight preparation, and how it seemed more "difficult for her on the ground than for us in orbit."

Skylab-4 astronauts celebrated Christmas by making a wire Christmas tree of scraps and rags they scrounged aboard their habitat. Celebrations like this seemed to ease the passing of time.

Some time during the middle of a flight, the spacefarers usually do what so many people in confinement or isolation do: they count the days to their "release."

During the long months of routine, depression sometimes sets in and sleeping disorders appear. Berezovoy and Lebedev were instructed to sleep two hours a night longer to dispel some of the tension they were experiencing. Hostility might express itself subtly. For instance, while taking a routine blood sample after a period of particularly heavy stress and hostility, one cosmonaut unconsciously (but was it entirely inadvertent?) stabbed his crewmate's finger so deeply that the wound became infected.

The Russians have a variety of measures to make cosmonauts feel as if they are still in the family of man and a species of Earth while they travel through the remote and cold void of space. They have conscientiously applied themselves to answering the question: What makes existence human?

For one thing, relationships with other people make us human. To satisfy this need they typically send visitors, in the form of visiting crews.

These visits are a mixed blessing. They are occasions to look forward to: "New people, new emotions," said Ryumin. "New information which we could not always get from [radio] communications." But visits are a nuisance too. If a foreigner is aboard, it means unwelcome public relations sessions every day, which take some preparation and interrupt the work. It means crowding. It means a lot less sleep because of the activity and excitement. Lebedev noted in his diary that it meant adjustments in one's relationships: "We have already settled down in our relationship, and now there are new people. . . . We two have got used to each other, worked it out, and now it looks as if we have to start again."

Later he elaborated: "Our attitude toward the visiting expeditions is complicated. On the one hand, I associate their arrival with a lot of commotion. We had to tidy up the station, then feed everyone. . . . On the other hand, you wait for them like brothers and the joy of personal contact makes up for the difficulties. . . . You receive great emotional support, of course, during the visiting expeditions, but get very tired as well. We slept an entire two days after they left."

Cargo ships also bring up supplies regularly. There are always treats and surprises stored away: letters, fresh fruit and vegetables, packages. In space's isolation and monotony there is a joy in the surprise, a high value placed on the new and unexpected, especially if it involves stimulation of the senses or some nourishment for their atrophied emotional life.

Foods can set off a long train of thoughts, memories, and emotions. "We found dried apricots," said Aleksandrov, "and it's as if it's closer to Earth life. But there is something that separates it from the usual, Earth way. Everything is remembered as if it was very long ago and we are not together and it is not known when it will be again."

Somebody stashed a copy of a nature book on a cargo ship going to Ryumin and Lyakhov, for which the cosmonauts were very grateful: "It was pleasant to turn the pages and once again to look at the running brooks, at the rivers and lakes, to remember the tie with the Earth, from that which we had left, but where we shall always strive to return. It is our home."

Lebedev found that he loved to read old newspapers, even if he'd read them many times before.

Containers of pine scent are welcomed. A videocassette of a child's birthday party and some Moscow street scenes were eagerly received and watched again and again.

For leisure activities cosmonauts enjoyed videocassettes of concerts and variety and comedy movies—110 of them. "There are never enough," said Aleksandrov. They listened to a wide variety of music too. "We were once sent a cassette recording of a rooster crowing, a mooing cow, flowing water, and other sounds," 140-day veteran Aleksandr Ivanchenkov later told an interviewer. "What a pleasure that was!"

As part of the psychological stimulation and support, cosmonauts talk to a wide variety of people outside the narrow world of Mission Control professionals—famous entertainers, interesting lecturers, sports commentators, and explorers. They were even briefed on the most recent SALT negotiations by people from the political indoctrination office. Such meetings are scheduled every few days during a long-term flight.

The optimum space voyage in terms of psychological health and work efficiency is considered to be about three months long. There is not enough time to accomplish long-term projects in less time, and it takes a few weeks to adapt to space and reach one's stride in terms of work. Beyond four months stresses increase, fatigue sets in, rest becomes disturbed, and even work efficiency begins to deteriorate.

The Lyakhov/Aleksandrov crew (150 days) and the Kovalyonok/Ivanchenkov crew (140 days) found that the beginning of the fourth month, at about day 100, is a critical point in a long expedition. That's when they usually experience a psychological decline, depression, nostalgia. To combat this effect mission planners add new experiments into the work program at that point. These evoke much interest on the part of the cosmonauts. More concerts are beamed up, and, as mentioned before, contacts with families become longer.

Nonetheless, the mental countdown to the end of the flight continues. "Only two months from now will I be able to think and hope about the landing," one wrote in his diary.

The inner life is enhanced for some by keeping diaries. These become more than a recording of the days' events—they are journeys into the soaring, spacefaring soul. "I wrote every day for the entire 211 day flight," recalled Lebedev. "I tried to write down my impressions in every free minute. . . . but mainly at the end of the working day. The diary became my comrade, so really there were three of us on the mission: Anatoliy, myself and the diary."

Dreams can be a vivid and pleasant means of escape, too. Grechko and Romanenko dreamed of woods, rivers, and going skiing. Lyakhov dreamed of going fishing in the early morning, watching "a fine dawn," and catching an enormous fish. Aleksandrov said that while he was on the station he dreamed of Earth differently than he did after he got back.

In a rare glimpse into the private soul of an American long-term spacefarer, Skylab commander Gerald Carr recently disclosed an entry from his diary. "Up until day 50, all my dreams were as if I were walking and moving in 1-g. But on day 50, I had my first zero-gravity dream, in which everything moved around weightlessly as it would in space. It was as if my subconscious let go of Earth."

Carr recalled being very uncomfortable and restless his first night back on Earth; he tossed and turned between one-g and zero-g dreams. On his first night back on Earth after his long flight, Aleksandrov dreamed of seeing Earth through a port-hole. Lyakhov too dreamed of being back aboard the Salyut on his first night back on Earth.

And afterward, too, the spacefarers sometimes revisit space in their dreams. "The station and the mission are starting to appear in dreams," Lyakhov told an interviewer, "while out there in orbit, dreams were usually about the Earth."

It would misrepresent the space experience to give the impression that there are no moments of ecstasy, that it is a routine of unrelieved gloom, work, and pent-up emotions. There are two sources of exaltation: one from the realization that many people are cheering you on from Earth and regard you as a hero, and the other from the sheer beauty of Earth and the stars, and the sensual splendor of flying weightless through space.

On some occasions the cosmonauts feel the public nature of their work as a source of pride rather than a burden and responsibility. "They read us a stack of congratulatory tele-grams from relatives, friends, acquaintances and from total non-acquaintances, and even from whole collectives," noted Ryumin. "We began to sense even more how many com-pletely different people were following our flight. I was even more convinced of this after the flight when I was told how many people in various parts of the country watched the progress of our flight on the television evening news. We owe them all a tremendous gratitude."

In the Soviet Union cosmonauts are still decorated with medals and honors for their work, and are held in the highest esteem by their nation. They are often treated as a combination of Hollywood stars and royalty; if martyred for the cause, they become saints of the new Soviet religion. People in the small towns in the countryside rush to meet them, if they hear that the cosmonauts are coming, and fill their arms with flowers. This adulation is important and appreciated; it assures them, even while they are in space, of their worth, the importance of their sacrifices.

And there is always Earth's beauty, a great relaxation and aesthetic escape, and a unique privilege unlike anything on the planet's surface. Spacewalks are especially splendid because the vista is unrestricted.

Aircraft pilots sometimes experience something called "breakoff," a wonderful feeling of freedom and rapturous detachment that has been identified by psychologists as a transcendent and altered state of consciousness. It may even have a physiological basis, and it is possible that a certain number of plane crashes caused by pilot error are really due to this phenomenon.

There has been a question of whether Romanenko experienced this sense of rapture when he almost lost his life during a spacewalk in December 1977. Some American psychologists came to believe that he deliberately flung himself out of the airlock in a state of ecstasy and literally heavenly rapture.

The real incident, however, was much more prosaic. During a break in the activities he decided to look out the hatch, but he missed his grip and floated right out. He had unlatched his safety line while working inside the airlock, and had not resecured it before making his move. Fortunately, his crewmate, Georgiy Grechko (who was doing the exterior work), was able to grab him as he floated by. This saved his life. (And the agitated cosmonauts decided not to tell Mission Control about the incident until well after the flight.)

Astronauts and cosmonauts have frequently spoken rapturously of the beauty of space, but have never said they experienced an altered state of consciousness or anything like pilot breakoff. Although American astronauts might keep it to themselves if they did experience it (as they regularly do with practically all their emotional responses), there is no reason to believe that the Russians would do so since they religiously share all their feelings with psychologists. This would be an especially interesting phenomenon to investigate, if it existed.

The last month of a long orbital flight has been called by Russian psychologist V. I. Myasnikov the "breakaway" phase. "During the 'last dash' period, the cosmonauts exhibited high motivation for work and very coordinated actions," he wrote.

Preparation for return to Earth, coincident with the last stage of confinement, is characterized by uneasiness, excitement, and, oddly enough, regret. For Lebedev, sleep came hard the night before his return. "A mood I do not understand: worry," he wrote in his diary. "How are things down there? Our life has been adapted to a small little island in space, and then suddenly, the Big World! I am not myself." Ryumin admitted that he and Lyakhov were emotionally tired. "Maybe we were saddened at the thought of parting with the station which had served us faithfully and honestly for six months. We felt as if we were parting with a good friend forever."

Berezovoy, too, was surprised by his regrets: "It was not as joyful as we had thought. I had mixed feelings. On the one hand was the desire to go home. On the other a sense of dissatisfaction: it seemed that on such a long mission more could have been done. Yes, during those seven months the station had become home."

American astronauts had expressed similar sentiments about leaving Skylab. "We were ready to go home," Pete Conrad wrote. "But it had been an extremely comfortable home while we'd been there, and we sort of hated to leave it, to get back to one g."

One difference between the American and Russian attitudes may, at least subconsciously, have been based on the astronauts' realization that the one-shot Skylab would be gone forever. Soviet long-duration cosmonauts have generally made repeat visits to their own space stations, so their departure from orbit is not a permanent good-bye, but just *dosvidanyah* *("au revoir")*.

Russians and Americans always tended to leave something for the spacefarers who came after them aboard the station. In the case of the Russians, it was usually a letter; in the case of the Americans, one time it was a dummy humorously propped up on the bicycle ergometer. But these traditions will fade when stations are permanently occupied, without any unmanned phases between visits. Then the crews can develop new traditions for handover and "change of shift."

Significant activities must precede this handover point. The relieving personnel conduct thorough inventories and status checking of all equipment and supplies, and any discrepancies are remedied or waived in writing. Missing and misplaced material is located. Traditionally, any members of the relief crew (including the commander) can refuse to accept handover until they are completely satisfied with the revealed status of the facility. This refusal probably must be a genuine threat, and without it, required checks and inventories may not be done with sufficiently attentive diligence—a possibly fatal error in space operations.

Once satisfied, individual crewmembers can proceed with duty handovers at their own pace. The commander's handover will be more symbolic; a key, a logbook, or some other traditional totem of orbital authority will physically change hands to an appropriate audio accompaniment.

The relieved crewmembers may not want to rush back to Earth immediately. There could be a day or two of "vacation" when, with all responsibility lifted, they can fully savor weightlessness and the view out the window. There will be personal items to pack (they will henceforth sleep in the docked shuttle since their apartments are occupied by the new crewmembers), good-byes to be made, final promises to be kept, last letters written for the special "spacemail postmark" widely prized earthside as a souvenir. The road back to Earth traverses many psychological as well as physical barriers, and the route must be negotiated as carefully as possible to ease the spacefarers' difficulties.

Return to Earth will be wonderful and awful in its own way. None of the cosmonauts could help noting the sounds, smells, and sights of Earth, as if sensory deprivation had made their senses ultra-sensitive, hungry again to embrace Earth living and its pleasant sensory bombardment. Lyakhov told a TASS correspondent, "My first impression was the odors of the steppe which rushed into the ship when the hatch was opened."

But the sensation of weight is unpleasant. The cosmonauts noted that a cup of tea sinks like a large meal to the pit of the stomach; a bunch of flowers feels as heavy as a sheaf of wheat. Earth's gravity clings heavily to all parts of the body. The returned spacefarers perspire heavily when they take a few tentative steps, and the medics frequently ask them to lie down. They obligingly and gratefully comply.

Back on Earth the spacefarers' emotional lives are revived. Welcomed enthusiastically by their rescuers and well-wishers who are ready to tear the spacecraft apart for souvenirs, they may pass out gifts or souvenirs. This is a very Russian tradition. Once they're back at the recovery hostel in Tyuratam, the phone will start ringing. These calls from home are joyous and rambling.

Sleeping on earthly beds can be a problem. "The most difficult thing for a cosmonaut is his first night on Earth," explained Aleksey Leonov. "Even a bed of down seems too hard after months of life in conditions of weightlessness."

Returning cosmonauts spend several days recuperating before they are allowed to go home and resume their normal lives. Lyakhov claimed that the longest day for him was the last, just before being reunited with his family. For the spouses too, telephone communication cannot take the place of physical, tactile contact, as if touching, not seeing or hearing, is the true measure of reality, the only genuine reunification. "I regained my peace of mind," said Alyona Romanenko, "only when I touched Yuri after he landed in Baikonur."

The chain that was broken at launch must be reforged after the emotional reunification. The final earthward step is reassimilation into the family, and it could take a long time. Psychologists point out that families that adapt badly to a person's absence integrate him or her back into their traditional roles and responsibilities quickly. Conversely, families that have adapted well to a person's absence and have spread the duties around often have some difficulty reabsorbing a returned member. The traditional role of, say, the head of the household, has been taken on by someone else. Decision-making and nurturing have been relinquished and are not easily given back. As Harrison and Connors have written, resentments might also have built up in the interim, on the part of the spouse who has had to take on the whole burden of family responsibilities. They would agree with Ryumin that the heaviest burden may be with the person who is left behind.

Some cosmonaut wives accept the traditional role of second fiddle, the wife of a great man. "If a wife sees that her husband is completely dedicated to a new undertaking, if he gets complete satisfaction from his work and if he is happy with it, can she protest, get upset and be an obstacle in his way?" said Svetlana Leonova. "Having a happy man around the house is a wonderful thing. It's a sign of your happiness together."

But for most of the women, a happy man around the house is not enough. "I'll never give up my work," said Natalya Ryumina, a computer specialist and engineer, "because being a cosmonaut's wife is not a profession." Galina Kizim agreed: "As for my job at the design office, I'll never give that up . . . I'm not the only wife who tries to combine a home and a job."

As in most American families in which both spouses work,

the women shoulder most of the burdens of housework and child care. "Aleskey is busier than I am, so naturally most of the household duties fall on me," said Larisa Yeliseyeva. "The fact that Aleskey doesn't care too much about what I feed him for lunch and dinner is a big help."

The long space flights are hard on closely knit couples with little children. When Yuri Romanenko went on his 96-day flight, his wife, Alyona, took it hard: "Even on Earth we had never been away from each other for so long. And there I was all alone with two children. Our younger child, Artyom, was just a few months old."

Svetlana Leonova, the older, more traditional wife, took the absences and the risk stoically, as the price one pays for being married to an interesting person: "We aren't the only wives who wait. The wives of seamen, the wives of polar explorers, geologists, and builders wait too. I don't go for talk like 'I was terribly afraid' or 'I was so worried.' Any new venture involves a certain amount of risk." Svetlana Leonova's stoic and lofty view must have served her well over the years, since her husband was in the first cosmonaut class, along with Yuri Gagarin. In those days, cosmonautics was much more dangerous than now, and some of Leonov's comrades were killed in training and spaceflight.

She willingly accepted the role of celebrity wife too (the family car is a large, grand, shiny Mercedes), though it could not have been an easy one since she was the Russian equivalent of the wife of an American "original seven" astronaut: "I'm always under public scrutiny. I have to be on my toes all the time and know when to move unobtrusively into the shadows."

But it is natural to be afraid or worried, and it isn't always easy to move into the shadows when two professional lives—and children—are at stake.

Less role-oriented couples, especially, in two-career families with children, must discuss things first. "It is a rule in the family not to interfere with each other's plans," said Larisa Yeliseyeva. "We always discuss our plans and decisions together. I'm always interested in Aleskey's opinion about what I intend to do, and I suppose he's just as interested in my plans, but we never force anything on each other."

The Russians also have an excellent support system of friends and relatives to help the cosmonaut's family in his or her absence. Many of them live with other cosmonaut families in Star Town where there is always sympathy and help. New and special friendships can spring up between the families of the absent cosmonauts, as one did between the wives of Popov and Ryumin (who had not even met before the launch). Even civilian cosmonauts, who do not live in Star Town but in a special suburb north of Moscow, speak of the large number of friends who surround their loved ones and are always there to welcome them back.

The cosmonauts and their families have that intangible quality that helps them endure the pressures of space work: courage. "In actuality, it is unknown what requires more courage," Oleg Gazenko told Russian journalist V. Yankulin, "to perform some courageous act or to once and for all stop smoking. Probably the most important thing is there is a deep awareness that you have no other recourse. There are no pills or hypnosis or injections—nothing can replace this awareness. Only it could force man, even on the Earth, to carry out all those procedures which the cosmonauts carried out in orbit. Every day, regardless of mood, of the current situation, regardless whether it is your birthday or your wife's, everything has to be done in full measure. And in actuality, it requires real human courage to every day carry out a great many tasks which are not always joyful, but are sometimes downright unpleasant, which require several hours a day. . . . Everything has to conform to the rigorous work schedule. And it is precisely this which constitutes the real heroism, which is given birth by the great internal mobilization of these people and their clear understanding of the goals set for them."

To the questions, Why am I here? What is the worth of my

life? the cosmonauts and, to a degree, their families answer: "The important work, the pioneering work that I do here exploring and exploiting space."

The spacefarer's life is never completely transferred into space; in practice only a small portion of it is. Until people can live with all their being in space, and embrace the full number and variety of roles that define their identity on Earth, they will always feel lonely to a degree, as if something vital were missing in the vast expanse of space.

Frontier literature tells us that women and children ultimately civilized the West. One woman who had been widowed on the road west set up a small house near a prospecting area where she offered her domestic services—took in laundry, offered clean rooms to boarders, and cooked meals. Men paid an exorbitant price just for the privilege of sleeping on her porch. Her value was not primarily in the services she offered—most of these men had been cooking for themselves and didn't care if they smelled like goats—but in what she represented: civilization, home, domesticity, comfort, everything so sadly lacking in the wilderness.

At the risk of sounding sexist, cosmonaut Svetlana Savitskaya had a somewhat similar effect on the first mixed space crew in 1982. Lebedev wrote, delightedly, that she had kept him and Berezovoy waiting while she fixed herself up: "When we opened the hatch after the docking, we expected that she would make her appearance from the vehicle, but like any woman she was preening herself and calmly saying, 'Now, now.' And then the tail of her hair [Savitskaya usually wears her hair in a pony tail] appeared in the hatch." Lebedev jokingly handed her an apron. She balked at it, saying that it was up to them, not her, to remember their manners and feed the hungry travelers, their orbital guests.

But most important of all was not their deliberate joking gesture but an unconscious one: for the first time in the entire long flight, they washed their hands before dinner.

There is a civilizing influence on space populations even

more important than women's, since the women in the current and near future programs have nearly identical credentials, attitudes, and backgrounds as the men. This genderless cultural similarity is so strong as to raise the question of whether they will ever constitute a truly heterogeneous group, or will be only a variant of a very homogeneous one. In any case, this stronger influence will be the presence of children, leading to the reorganization of some pockets of spacefaring life along nurturing and domestic lines.

Consciously and unconsciously the cosmonauts found their wives, their children, and the normalcy of their domestic lives the things they missed most in space. Ultimately, while the Russians extend the duration of their spaceflights, they do so cautiously, perhaps questioning just how far a person can be pushed before he finds this uprooted life unendurable. "The limitations to living in space are not medical," said Gazenko, "but psychological."

Psychologists have found that in isolation and confinement studies that small, homogeneous groups tend to opt for passive leisure activities and reduced interaction. However, large, mixed groups show much greater interaction and creative use of leisure time. "Apparently in these experiments it was the women who instigated group activities, perhaps through a felt need to entertain the children," wrote Harrison, Connors, and Akins. "The high level of interaction among these confinees is very provocative and suggests that heterogeneous groups, especially those including women and children, may have a completely different approach to the use of leisure than groups of young men, the usual subjects of confinement studies."

In remote places, leisure activity becomes varied, and life becomes active and enriched in a mixed community of men and women, young and old. These are symptoms of the human soul's revival. Space will be such a place when people can be with their loved ones, and go on with those fundamental relationships from which so much of our identity emanates and without which we do not thrive. It will have to become a place where the spacefarers are not left, as they are now, dangling in emotional and spiritual limbo, waiting for the return to life, to Earth.

These returning spacefarers may thus experience an emotion not unlike that expressed so poetically by Charles Lindbergh more than a half a century ago. In describing his feelings when he spotted the Irish coast after his tortuous, pioneering, solo flight, he wrote: "There is Earth again, the Earth where I've lived and now will live once more . . . I've been to eternity and back. I know how the dead would feel to live again."

chapter seventeen

— The —
COMING
SCHISM

Their readiness to achieve sociopolitical and resource independence will grow with the psychology of their extraterrestrial motivations and their technological ability to create new worlds in their totality. This will lead to [their becoming] not a colony of Earth, but a sovereign, mobile neocosm. . . . In [this] phase, we leave the harbor and emerge into the open sea of space, psychologically and socially. Human history henceforth will pulse through many world-arteries that lose themselves beyond the horizons of our present perception in the trackless infinity of space and time. . . . The human life form may be regarded as returning to the three-dimensional origin of all terrestrial life. The two-dimensional existence of Earth's land surface becomes an evolutionary benchmark wedge between the three-dimensionality of the finite oceanic womb from which life rose to the brightness of consciousness and the infinite cosmic womb in which it can rise to a level beyond our understanding.—*Krafft Ehricke (1917–84)*

Dr. William G. Niederlander once envisioned a new branch of psychology establishing itself in the future: "psychogeography," the study of the psychological separation of the species from Earth.

In fact, the psychological separation began with Konstantin Tsiolkovskiy and Robert Goddard in the first years of this century. "The Earth is the cradle of the mind," the Russian wrote, "but one does not live in the cradle forever." Tsiolkovskiy was then a mathematics teacher at a fashionable school for girls in Kaluga, a hundred miles from Moscow. Between classes and faculty responsibilities, he dreamed and wrote not only of spaceflight, but of the establishment of space cities and a space civilization. Meanwhile, in Massachusetts, Goddard was beginning to design the ancestors of true spaceships.

Jules Verne, H. G. Wells, Wernher von Braun, Hermann Oberth, and Krafft Ehricke, along with Goddard and Tsiolkovskiy, have all taken that journey which precedes physical exploration, what historian Daniel Boorstin called "an adventure of the mind, a thrust of someone's imagination."

Today, many hundreds—even as many as a few thousand—earthers have no mental barrier to envisioning the day when we will travel between the stars or harness the physical powers of the universe to the degree that we can terraform, rebuild other planets into Earth's image. They are, like Henry the Navigator of fifteenth-century Portugal, standing on the shore of a sea they will never travel. But they are developing the tools, techniques, the ideas and the means by which others will.

In his last great speech before his death in 1984, the great spacecraft engineer and spaceflight philosopher Krafft Ehricke said that the "information metabolism" of current civilization was the greatest technological advance since photosynthesis. This is so because, like plants, we take in information, decompose it into abstractions and generalizations, and resynthesize it into life. The space age has brought us into what Ehricke called an "umbilical stage of metabolism which interacts between the biosphere of Earth and the ectopic wilderness on the outside."

Psychogeography, for Ehricke, is already here. "Information cannot be recycled generation after generation, because it uses all the elements of Earth and beyond. We are destined to go to other spheres by it."

In Earth's history, the oceans were barriers to some nations and highways to others. There was a demonstrably intimate relationship between such attitudes and the nation's prosperity and intellectual vigor. Space now plays that role—to be a barrier or a network of highways. We have the maps of the mind, and we have the spaceships and the spacefarers, and for the moment we have the will. We cannot turn our back on space as China once turned its back to the sea without suffering, as the Chinese did, a stagnated civilization turned in upon itself.

Many people have speculated on what human beings will be like as a result of what Ehricke called "the extraterrestrialization of our species." Ehricke himself thought the Moon would be a good proving ground over the next fifty years to experiment with settling space. "Technology will make us fit to live in the Solar System," he said, "to fit into the ecological niche of space. Some people used to say, if God meant people to fly, they would have wings. Now the joke goes, if God meant people to be a spacefaring species, he would have given them the Moon!"

Developing the tools to assault the unknown is far from a straightforward process. Predicting human development is even riskier than predicting technological progress. But, based on the past, Ehricke believed that the "divergences of sociological repercussions of space living on Earth might have an impact similar to that which the United States has had on Europe in the past two hundred years."

It is possible that the people who settle in space in the next few decades may not change at all. They may return from space with the same views, prejudices, and limitations they had when they left. This is especially likely for people whose exposure to outer space is short, a week or a month. Vast, profound changes in human beings take time, sometimes even generations.

Skylab-4 commander Gerald Carr, who lived aloft for nearly three months, found that "Being out there washes away notions of nationalism and provincialism, gives you a more universal feeling. I had to watch carefully to see anything man did on the planet, and it makes you feel a little insignificant. I came back with a different attitude toward the environment, toward the universality of man, toward life."

It's possible that with the permanent settlers in space, some changes, both subtle and profound, may be mandatory in order to ensure survival. The unforgiving environment of outer space will almost certainly teach its own virtues, the most important of which may be that total self-reliance and individual independence may have to be sacrificed for the survival of the group. This has been common to all frontier societies.

Within any society, the size of the community influences how the individual will behave. It has almost been taken for granted that the size of space settlements will evolve in linear fashion, from small to large. For several reasons, this might be the most misleading fallacy of our current thinking about the future. Although it's possible that we could scale up to space cities of tens of thousands, as envisioned by Dandridge Cole and Gerard O'Neill, there are many practical reasons that such large-scale colonies may be undesirable for at least the next hundred years. As NASA spacecraft designer Maynard Dalton explained: "It's just not easy to scale up power systems, life support systems, and waste removal."

Many scientific and industrial processes may preclude building a large space station in favor of several smaller ones. Diverse station activities tend to conflict with one another. In the early 1980s, the cosmonauts on Salyut-7 had to shut down many systems in order to run the energy-demanding crystal-growing furnaces. On future space stations, telescopes and even some industrial equipment can be jiggled so much by people bouncing around a station that their data can be rendered useless. Already, some space telescopes are built to be undisturbed free-fliers. Earth resources scientists may need a station (or several) in polar orbit, whereas other disciplines will prefer some other orbital paths. As a result, there may be an early need for various endeavors to possess their own dedicated space stations. They're cheaper and easier to build than large space cities, and can accommodate specialized needs more harmoniously.

Dalton predicted that the future development of space will

probably involve many small colonies (ranging in population from ten to one hundred, averaging about twenty) rather than the establishment of one large space city. In essence, space development is unlikely to be homogeneous, simple, and linear, but instead diverse and complex.

The only really important reason to have a large settlement in space is if it is located at an important crossroads, just as the cities of the western frontier were established in the last century. It is quite possible that in the next two decades many small manned and unmanned free-flying platforms—astronomy platforms, Earth-observation satellites, pharmaceutical and crystal-growing factories—may cluster around the main American space station. The space shuttles will, after all, be visiting the station at predictable intervals, and it will be cheaper and more convenient for everybody if the shuttle can rendezvous with several structures in the same neighborhood. If such places also become import/export centers to which Moon, Mars, and asteroid miners can bring their resources in exchange for supplies they require, then there may be a need for a large space city. If entertainment, meetings, and political summits are also located there, it may become a large colony of a few thousand, some of them merchants and tourists from Earth, others off-world traders and visitors.

Also, large space towns would be necessary to house construction crews building gigantic space structures, such as the space solar power satellite, an object many miles on each side and requiring a construction crew of a thousand. At least that was the rationale ten years ago. Economics, however, might dictate deploying smaller, unmanned mirrors, or placing such power stations on the Moon. This might be a stable platform for which the Moon's own resources would provide the building materials. Such systems require much smaller initial investment, and show more rapid return, and they have the potential for scaling up—all attractive alternatives which make such a big project unlikely.

Planetary development in which mining and scientific interests can be scattered all over a planet would also argue for the establishment of many small settlements and one somewhat larger supply depot where interplanetary spaceships can leave cargo. The only reason to have a large centralized city-type region on planets is as a supply base for interplanetary goods, to which settlers from remote regions can go for the items they cannot obtain or grow themselves.

Because the expansion into space will involve many different forms of endeavor, regionalisms may develop according to the needs and tasks associated with each distinct group. For instance, the attitudes and rules of business-driven industrial parks may differ from military and security-minded forts, which will be different from purely scientific laboratories and observatories. Early on, of course, all these groups will be jumbled together on one space station, but it's not hard to predict that once outer space becomes easier and cheaper to develop, specialized space stations can be established to accomodate common-theme activities.

If, say, there are groups of miners on the Moon, scientific researchers on Mars, crystal-growing factories orbiting Earth, and a pharmaceutical house established in outer space, it's possible that these populations might meet only on an orbiting casino/hotel/amusement park complex. Isolation, communality of activity and aim would certainly be important factors in developing distinct, almost regional, characteristics, just as some of the geographical features of New England, the South, and the Mid-Atlantic states shaped some of the early American settlers.

The size of a colony is important since it dictates, to a degree, how the individual relates to society. For one thing, individuals have more impact on the culture of a small colony than a large one; in essence, one person's contributions matter more in a town of forty than in one of a thousand.

Paradoxically, balanced against the individual's greater impact in a small group is more pressure for the individual to conform to the group once a path is established, especially if the group's survival is threatened. The outcome depends on the group. In his brilliant book *The Last Kings of Thule*, explorer Jean Malaurie described the Eskimo culture of a generation ago. He noted that in Eskimo groups living on the fringe of survival, "the group thinks. One's personal ideas are suppressed. . . . [T]o behave otherwise would have seemed incongruous, inviting ridicule." The Eskimo wore a series of social masks, as if living in such a tightly knit group required concealing one's essential nature. Nonconformity could be accepted only in the form of shaman or buffoon.

Furthermore, under such threatening conditions, the Eskimo society's weak individuals were apt to be sacrificed to the elements, while the hunters and producers of the society, and their offspring, were favored. Old people were treated with contempt, and were encouraged to commit suicide. Sickly or deformed infants were killed outright, and orphans were badly treated.

It is possible that a Martian colony, technologically advanced though it may be, may, like the Eskimos, have limited food supplies and a delicately balanced ecosystem with little room for the superfluous. People with hopelessly terminal illnesses may not have the luxury of staying alive indefinitely in a debilitated condition or on life support as they do on the somewhat more abundant Earth. Deformed children—and there could be a few more of them due to environmental factors beyond Earth—might also not be supported.

Thus, the "psychogeography" of the space frontier may start with a return to less highly evolved social practices dictated by such an environment.

The American Puritans, the early Quakers, and the Virginia Episcopalians had their own variants of "group-think," although the individual had somewhat more to say. This was also the congregationalist tradition, in which all opinions could be taken into account, and from which our contemporary political institutions to some degree sprang. Group-think could lead to the suppression of certain chronically nonconformist individuals, and an unwillingness to open up to ideas from outside the community. In remote, rural Virginia, the Episcopalian religion had to constantly strive to accommodate and meet the current needs of the community. In all early settlements, compromises had to be reached—either the individual bent to the society or the society to the needs of many individuals. The more people, the more need for compromise, and a tendency to water down the purity of rigid principles.

Krafft Ehricke believed that societies first became pluralistic, and then united as a civilization "to reach an ascendancy beyond plurality, a seeking of values by which to live, like natural laws." Once human beings go beyond the "umbilical stage" we are in now, how will they develop? Will they be like us, or will they be different?

For one thing, they will look different if artificial Earth-normal gravity is not maintained throughout the Solar System. Earthers might be able to tell at a glance a "spacer's" planet of origin by looking at the shape and development of his or her body. To return to Earth, "spacers" will suffer an incredible form of interplanetary jet lag (or "spaceship lag"), a long recuperation on Earth's surface until mobility would not be life-threatening.

Beyond that, it is likely that the off-worlders will develop their own gestures, mannerisms, and speech characteristics. During the two world wars, counterespionage agents could often spot a spy by the manner in which he smoked a cigarette or held a fork. Such telltale signs will pop up in space natives. Their gestures might be slightly exaggerated, like those of a mime, to compensate for the difficulty of expressing oneself nonverbally in weightlessness. By the same token, they may not be aware of the oddness of their facial expressions since they are accustomed to their own facial messaging, developed to counter the natural distortions in the face. The way they cut their meat, spoon up their soup, hold tools, write with a pen, and all the other unconscious motions will be points of difference.

Of course, they might think differently too, according to their outer space niches and experiences. Those who live as interplanetary gypsies, or on supply posts between planets, will make the most profound leap of all, since they will be creatures of motion, not of fixed position. Pre-spaceflight maps of outer space (and even space age maps composed by people still locked into prespaceflight thinking) show planets pinned to the black star-studded background like dead butterflies on purple velvet. And these would-be cartographic representations possess as much fidelity to reality as do the dead lepidoptera specimens to the visual melody of a living butterfly in flight.

To understand the essence of butterflies, watch them fly. To comprehend the realities of space geography, keep your eye on the motion. Relative motion, not relative location, defines the true geography of space. Movement, not coordinates in any reference frame nailed to a physical anchor, is the quintessence of "locational identity." No sign posts or median strips lie along these spaceways. But the routes, truck stops, and speed traps are already roughly defined. Spacefarers will come to instinctively know all these unearthly rules of thumb and the uncommon, cosmic sense of such motion.

Their environment is likely to make these interplanetary gypsies the ideal celestial navigators. They may not be able to make that psychic leap overnight, and may require a home port somewhere, some fixed location of spawning and burial. Eventually, they may make a virtue of rootlessness by celebrating the freedom of motion. Their philosophy may be the only one approaching the "universal." Possible breakthroughs in thought and technology may be made by these people because of less rigid frameworks of time and space which their environment places on them. Or the opposite could be true: they could erect rigid frames of reference on themselves to counteract the constantly shifting reference points in their lives.

Planetary settlements, at first on the Moon and Mars, and later on the asteroids, the moons of Jupiter and Saturn, and so on, might be psychologically more comfortable, and may provide a second space culture. Psychological rootedness to some place may be an inherent necessity of our species.

How will these off-worlders, these "spacers," think of Earth?

They may love it as we do, for its self-evident beauty, its myriad sensual appeals, its inexhaustible variability. Some people, especially those who migrated from Earth to a space settlement, may feel real nostalgia and homesickness. We already know that space living is a life apart from terrestrial life, and the latter almost appears as a dream of happiness when one is living far from loved ones in a place with no fixed coordinates. Even once people are born in space and families are established there, some may still yearn for Earth and their cultural roots, as some early American settlers yearned for England and thought of themselves as Englishmen. Second-generation spacefarers may look to the countries of their grandparents' origin in the same way second-generation American Greeks, Swedes, or French look toward Greece, Sweden, and France—nice places to look up one's roots, to chat with distant relatives, and to visit, but nothing to bleed and die over.

Some spacers may hate Earth. Even for those in the best of shape, every movement is weighted as if one is attached to a ball and chain. There is no freedom of movement, and it takes fifteen hard, exhausting steps to get as far as one effortless hop on, say, the Moon. A glass of iced tea will enter the stomach like Thanksgiving dinner. The air will be humid and foul-smelling to the Moon native. And the planet is, true to reputation, teeming with life—insect life. Imagine a world free of mosquitoes, gnats, and cockroaches, and then imagine coming back to Earth where these intrepid little vectors of disease and filth thrive. If this Moon native travels into any large city, he or she will see congestion, pollution, and traffic, and will hear noise and confusion. Hotel rooms are cluttered with furniture, and no Earth mattress can ever be as comfortable as a sleeping bag in weightlessness. The Moon native may pass out as the blood pools, from time to time, in his or her shaky legs. At vacation's end, he or she may well ask, "How do people LIVE here?"

It is hard to imagine how people can be happy on any other planet but Earth when we see the arid wastes of the Moon and the red dust of Mars in photos. When we first saw that splendid picture of Earth, an opalescent blue and green jewel on the black velvet background of space, we were all reminded how much we love it and want to protect it. But we are an umbilical civilization which hasn't yet left the Earth cradle emotionally. Some future "spacer" will undoubtedly lecture us on our provincial attitude. Who is to say that someone living on a moon of Saturn, overlooking the rings, or in a Martian building on the peak of Olympus Mons, or in a domed house along Hadley Rille on the Moon, or floating freely beyond Pluto surrounded on all sides by stars and comets, hasn't found the perfect place in the Solar System in which to live and raise the next generation?

Perceptual changes have always preceded political changes, and they are likely to continue to do so. Before long, populations who live off Earth will have to develop their own codes of justice, their own laws, their own courts, and their own punishments. It is impractical to haul every case back to Earth for adjudication. And why bother? What politically appointed Earth judge could grasp the essential nature of disputes in space?

Local justice could be administered by a hierarchy, or by the entire group if the local population is small enough. This has been the frontier tradition. Or if legal matters arise very rarely, judges could be sent out from Earth, and ride the circuit as they did in colonial times. Or disputees could be brought to some centrally located spot on specified court days, except when distances are simply too great.

It is even possible to hold court through telecommunications, like a long and large conference call. In any case, off-worlders might find it offensive to have their cases decided on the basis of the opinion of twelve good, ignorant earthers and true or, even worse, some international gaggle of earthers who cannot even agree among themselves on the important bases of the law. No, better to establish laws, courts, and juries off-world to deal with off-world problems, rather than face the absolutely certain problems of trying to interface closely with the already contradictory, arcane, time-consuming, and ponderous mechanisms of justice on Earth.

Once space societies become more self-sufficient, independent, and accustomed to ruling themselves, they may find they have become their own nations.

When the subject comes to money, the mother planet may wish to bring the pioneers back into the nest. It might reach out and, at some point, try to strangle off-world economics and the pursuit of happiness in space with taxation, quotas, and embargos. In this generation and the next, off-Earth industries will first be taxed by the countries who have reached out toward space. The United Nations has already made a power play of naked greed (i.e., traditional earther politics) by decreeing extraterrestrial resources "the common property of all mankind" (read, "under U.N. stewardship"), subject to taxes to the world body. Once space settlements establish their own legal status—which we believe will and should happen in our lifetime—it should be possible to have off-world corporations in which taxation benefits the off-worlders themselves, not the earthers who are its customers.

In spite of their probable international make-up, permanent populations of off-worlders will sooner or later become identified with one another rather than with their nations of origin. Their interests will be the interests of off-worlders, not of Americans, Russians, Japanese, Canadians, French, English, Australians, Germans, Brazilians, Saudis, Singaporeans, and so on. Shaped by the demands of a remorseless environment, bound together by unique off-world economics, formed by their own institutions and nascent cultural milieu, they will be a sovereign nation in mind and spirit even before they receive legal status as such. By nature they remain homo sapiens, but by nurture, homo extraterrestris.

This will be bad news for earthers. It is always easier to deal with a colony than a sovereign nation: colonies share institutions and are relatively pliant to the interests of the mother nation; sovereign nations are by definition perversely independent and, worse still, costly to deal with.

But independence is certain at some time, probably sooner than later. It is already accepted in space symposia that the farther from Earth space pioneers get, the more autonomous they and their technology must become. The sheer logistical problems involved in getting vital technology from Earth to some outpost in space are enormous. And communications between, say, Mars and Earth, will always have a five-minute to one-hour time lag, unless a future Nobel Prize winner in physics figures out some way by which long distance calls can exceed the speed of light. Settlers in the Asteroid Belt and beyond will feel even more distant. Dialogue will not be any more possible than it was for colonial Americans and the British Parliament. "Belter" speeches of independence may be like those of the American Revolution, long, eloquent, carefully reasoned documents of principle, the revival of a brilliant literary form.

In future political and economic disputes, huge populations on Mars, the asteroids, the moons of Saturn and Jupiter, and all the stopping stations in between will not take their marching orders from the beautiful but by then politically insignificant third planet of the Solar System. A common scenario among futurists is one in which Earth nations may be perverse enough to threaten vital exports like life-support technology, and off-worlders, especially Martians and asteroid miners, retaliate by cutting off by now vital raw resources. A war of independence could follow.

Independence has usually been won that way in the past, and both science fiction writers and futurists like to use the yardstick of the past to measure the future. But that scenario presupposes that our descendants will be squabblers and rebels like us.

They may not be. In this, the yardstick of the past may fail altogether, as it has many times before. Our descendants in space may escape the limits of our own imagination. The wonderful "extraterrestrials" of cinema and literature, with a civilization far more sophisticated and humane than ours, may never be strange creatures from beyond the Solar System; they might be our own great-great-grandchildren.

We in the twentieth century can only close our eyes on a starry night and try to dream of them, our shining descendants. Like Carl Jung, the eloquent troubadour of the collective unconscious, we travel to a house with vaguely classic lines, and come upon a dreamer in the innermost room. We peer into his face, and recognize our own. And we know that when the dreamer awakes, far in the future, we will be gone.

To the off-worlders we will never know, but whom we seek to touch in that house of vaguely classic lines, we say: Be free. Make your own life among the stars. We can think of no greater good for humanity than that.

appendix one

SPACE STATION MISSIONS 1971-85

Major Space Station Missions 1971–85

Mission name crewmembers, mission highlights	Date	Duration (days)
Salyut-1/Soyuz-11 Dobrovolskiy, Volkov, Patsayev Crew died on return to Earth	1971 June 6–29	23
Skylab-1/2 Conrad, Kerwin, Weitz Major repair operations	1973 May 25–June 22	28
Skylab-1/3 Bean, Garriott, Lousma Solar and medical observations	1973 July 28– September 25	59
Skylab-1/4 Carr, Gibson, Pogue Longest US flight; observed comet	1973 November 16– February 8	84
Salyut-3/Soyuz-14 Popovich, Artyukhin First sucessful Soviet mission: military version	1974 July 3–19	16
Salyut-4/Soyuz-17 Gubarev, Grechko First civilian Soviet space station	1975 January 10– February 9	30
Salyut-4/Soyuz-18 Klimuk, Sevastyanov Simultaneous flight with Apollo-Soyuz	1975 May 24–July 26	63

Major Space Station Missions 1971–1985

Mission name crewmembers, mission highlights	Date	Duration (days)
Salyut-5/Soyuz-21 Volynov, Zholobov Military vehicle, evacuated due to air failure	1976 July 6–August 24	49
Salyut-5/Soyuz-24 Gorbatko, Glazkov Repair effort fails? Station abandoned.	1977 February 7–25	18
Salyut-6/Soyuz-26/27 Romanenko, Grechko First two-port Soviet space station	1977 December 10–March 16	96
Salyut-6/Soyuz-29/31 Kovalyonok, Ivanchenkov Two guest-cosmonaut visits	1978 June 15–November 2	140
Salyut-6/Soyuz-32/34 Lyakhov, Ryumin Six months without visits	1979 February 25–August 19	175
Salyut-6/Soyuz-35/36/37 Popov, Ryumin Back-to-back mission	1980 April 9–October 11	185
Salyut-6/Soyuz-T3 Kizim, Makarov, Strekalov Repair mission	1980 November 27–December 10	13
Salyut-6/Soyuz-T4 Kovalyonok, Savinykh Final visit to Salyut-6	1981 March 12–May 26	75
Salyut-7/Soyuz-T5/T7 Berezovoy, Lebedev New space station	1982 May 13–December 10	211
Salyut-7/Soyuz-T9 Lyakhov, Aleksandrov Backup mission, nearly stranded	1983 June 27–November 23	150
Spacelab-1 Young, Shaw, Garriott, Parker, Lichtenberg, Merbold First Spacelab mission	1983 November 30–December 9	10

Major Space Station Missions 1971–1985

Mission name crewmembers, mission highlights	Date	Duration (days)
Salyut-7/Soyuz-T10/11 Kizim, Solovyov, Atkov Major spacewalk repairs; first USSR scientist	1984 February 8– October 2	237
Spacelab-3 Overmyer, Gregory, Lind, Thagard, Thornton, Wang, van den Berg Materials processing	1985 April 29–May 7	7
Salyut-7/Soyuz T-13 Dzhanibekov, Savinykh Repair and reopen three-year-old space station	1985 June 6–???	??
Spacelab-2 Fullerton, Bridges, Musgrave, England, Henize, Acton, Bartoe Astrophysics observations	1985 July 29–Aug 6	8

Spacelab D-1
Hartsfield, Nagel, Buchli, Bluford,
    Dunbar, Ockels, Messerschmid,
    Furrer
Materials processing
	1985 October	7

*Dzhanibekov completed his latest visit after 110 days on board; Savinykh remained on board with two new cosmonauts, Vladimir Vasyutin and Aleksandr Volkov, both pilots. They had reached the station aboard Soyuz T-14 in September, along with engineer Georgiy Grechko (who spent 8 more days aboard the station). The mission conducted the first space station crew exchange and command handover.

appendix two

SPACEFARERS
— and —
OTHER
EXPERTS

The following men and women have lived aboard "space stations" or space laboratories/workshops/observatories. The total duration of stay, even for very short visits, is given. Many of these spacefarers have been on other space missions not accounted here.

Name	Profession	Nationality	Days on board	Year of birth	Year of flights
Acton, Loren	scientist	USA	8	1936	1985
Aksyonov, Vladimir	engineer	USSR	3	1935	1980
Aleksandrov, Aleks.	engineer	USSR	149	1943	1983
Artyukhin, Yuriy	engineer	USSR	15	1930	1974
Atkov, Oleg	scientist	USSR	236	1949	1984
Bartoe, John-David	scientist	USA	8	1944	1985
Bean, Alan	pilot	USA	59	1932	1973
Berezovoy, Anatoliy	pilot	USSR	210	1942	1982
Bluford, Guy	engineer	USA	7	1942	1985
Bridges, Roy	pilot	USA	8	1943	1985
Buchli, James	engineer	USA	7	1945	1985
Bykovskiy, Valeriy	pilot	USSR	7	1934	1978
Carr, Gerald	pilot	USA	84	1932	1973
Chretien, Jean-Loup	pilot	France	7	1938	1982
Conrad, Charles	pilot	USA	28	1930	1973
Dobrovolskiy, Geor.	pilot	USSR	23	1928	1971 *
Dunbar, Bonnie	scientist	USA	7	1949	1975
Dzhanibekov, Vlad.	pilot	USSR	30+	1942	1978, '80, '2, '4, '5
England, Tony	scientist	USA	8	1942	1985

Name	Profession	Nationality	Days on board	Year of birth	Year of flights
Farkas, Bertalan	pilot	Hungary	7	1949	1980
Fullerton, Gordon	pilot	USA	8	1937	1985
Furrer, Reinhard	scientist	Germ'y, W	7	1940	1985
Garriott, Owen	scientist	USA	69	1930	1973, '83
Gibson, Edward	scientist	USA	84	1936	1973
Glazkov, Yuri	engineer	USSR	17	1939	1977
Gorbatko, Viktor	pilot	USSR	24	1934	1977, '80
Grechko, Georgiy	engineer	USSR	124	1931	1975, '77
Gregory, Fred	pilot	USA	7	1941	1985
Gubarev, Vladimir	pilot	USSR	36	1931	1975, '78
Gurragcha, Judger.	engineer	Mongolia	7	1947	1981
Hartsfield, Hank	pilot	USA	7	1933	1985
Henize, Karl	scientist	USA	8	1927	1985
Hermaszewski, Miro.	pilot	Poland	7	1941	1978
Ivanchenkov, Aleks.	engineer	USSR	146	1940	1978, '82
Jaehn, Sigmund	pilot	Germ'y, E	7	1937	1978
Kerwin, Joseph	scientist	USA	28	1932	1973
Kizim, Leonid	pilot	USSR	248	1941	1980, '84
Klimuk, Pyotr	pilot	USSR	69	1942	1975, '78
Kovalyonok, Vlad.	pilot	USSR	213	1942	1978, '81
Kubasov, Valeriy	engineer	USSR	7	1935	1980
Lebedev, Valentin	engineer	USSR	210	1942	1982
Lichtenberg, Byron	scientist	USA	10	1948	1983
Lind, Don	scientist	USA	7	1930	1985
Lousma, Jack	pilot	USA	59	1936	1973
Lyakhov, Vladimir	pilot	USSR	323	1941	1979, '83
Makarov, Oleg	engineer	USSR	18	1933	1978, '80
Malyshev, Yuriy	pilot	USSR	10	1941	1980, '83
Merbold, Ulf	scientist	Germ'y, W	10	1941	1983
Messerschmid, Ernst	scientist	Germ'y, W	7	1945	1985
Musgrave, Story	scientist	USA	8	1935	1985
Nagel, Steven	pilot	USA	7	1946	1985
Ockels, Wubbo	scientist	Neth.	7	1946	1985
Overmyer, Robert	pilot	USA	7	1936	1985
Parker, Robert	scientist	USA	10	1936	1983
Patsayev, Viktor	engineer	USSR	23	1933	1971 *
Pham Tuan	pilot	Vietnam	7	1947	1980
Pogue, William	pilot	USA	84	1930	1973
Popov, Leonid	pilot	USSR	198	1945	1980, '81, '82

Name	Profession	Nationality	Days on board	Year of birth	Year of flights
Popovich, Pavel	pilot	USSR	15	1930	1974
Prunariu, Dumitru	pilot	Romania	7	1952	1981
Remek, Vladimir	pilot	Czecho.	7	1948	1978
Romanenko, Yuriy	pilot	USSR	102	1944	1977, '80
Ryumin, Valeriy	engineer	USSR	358	1939	1979, '80
Savinykh, Viktor	engineer	USSR	74+	1940	1981, '85
Savitskaya, Svetlana	pilot	USSR	18	1948	1982, '84
Serebrov, Aleksandr	engineer	USSR	7	1944	1982
Sevastyanov, Vitaliy	engineer	USSR	62	1935	1975
Sharma, Rakesh	pilot	India	7	1950	1984
Shaw, Brewster	pilot	USA	10	1945	1983
Solovyov, Vladimir	engineer	USSR	236	1946	1984
Strekalov, Gennadiy	engineer	USSR	19	1940	1980, '84
Tamayo, Arnaldo	pilot	Cuba	7	1942	1980
Thagard, Norman	scientist	USA	7	1943	1985
Thornton, William	scientist	USA	7	1929	1985
van den Berg, Lod.	scientist	USA/Neth	7	1932	1985
Volk, Igor	pilot	USSR	11	1937	1984
Volkov, Vadim	engineer	USSR	23	1935	1971 *
Wang, Taylor	scientist	USA/China	7	1940	1985
Weitz, Paul	pilot	USA	28	1932	1973
Young, John	pilot	USA	10	1930	1983

A total of 34 Soviets (only 1 of whom was a scientist) have been aboard space stations, most (but by no means all) for very long stays. The Soviets have accumulated more than 3300 man-days of time aboard such facilities.

A total of 32 Americans (including 16 scientists) have been aboard space laboratories, most (but not all) for short stays. The Americans have accumulated less than 700 man-days of time aboard such facilities.

Fourteen other nationals have spent a total of about 100 man-days aboard orbital laboratories.

Note: Through 1985, assumes Spacelab D-1 flown as planned, does not count Soviet mission time after June.

*Died during mission.

references

Periodicals regularly scanned

Aerospace America (formerly *Astronautics & Aeronautics*), journal of the American Institute of Aeronautics and Astronautics

Aviation Week

Foreign Broadcast Information Service (Soviet Union)

Joint Publications Research Service (Soviet Union)

L-5 News

Space World

Spaceflight

Books and magazine articles

Adelman, S., and B. Adelman, *Bound for the Stars*, Prentice-Hall, Englewood Cliffs, N.J., 1981.

Akin, D. L., M. L. Minsky, E. D. Thiel, and C. R. Kurtzman, *Space Applications of Automation, Robotics, and Machine Intelligence Systems (ARAMIS), Volume 1: Telepresence Technology Base Development*, NASA Contractor Report 3734, October, 1983.

Akin, D. L., M. L. Minsky, E. D. Thiel, and C. R. Kurtzman, *Space Applications of Automation, Robotics and Machine Intelligence Systems (ARAMIS), Volume 2: Telepresence Project Applications*, NASA Contractor Report 3735, October, 1983.

Akin, D. L., M. L. Minsky, E.D. Thiel, and C.R. Kurtzman, *Space Applications of Automation, Robotics and Machine Intelligence Systems (ARAMIS), Volume 3: Executive Summary*, NASA Contractor Report 3736, October 1983.

Aldrin, Edwin "Buzz," with Wayne Warga, *Return to Earth*, Random House, New York, N.Y., 1973.

Aleksandrov, Aleksandr, "Earth, Wait for Us!—Excerpts from his In-Flight Diary," *Izvestiya*, February 11, 1984, page 3.

— References —

Allen, Joseph, and Thomas O'Toole, "Joe's Odyssey," *OMNI*, June, 1983, pages 60-63, 114-116.

Anaejionu, Paul, Nathan C. Goldman, and Phillip J. Meeks, *Space and Society*, American Astronautical Society, Univelt Incorporated, San Diego, Cal., 1984.

Anonymous, "Autonomous Spacecraft: Thinking Machines in Space," *Vectors*, No. 3, 1983.

Anonymous, "Cosmonauts Report Experiments, Meteor Shower," TASS, Moscow, July 27, 1983.

Anonymous, *Global Change: Physical Processes in Global Biogeochemical Cycles: An Assessment of Proposed Programs*, September 30, 1983.

Anonymous, "Perhaps The Hardest Part is Waiting," in *Soviet Life*, April, 1981, pages 42-45.

Anonymous, "Space Jokes," *Soviet Life*, April, 1981.

Anonymous, "Space Program Tester Tells of Sensory-Deprived Closed Life-Support Experiments," *Sovetskaya Rossiya*, No. 81, April, 1983.

Barchukov, I., "Overcoming Weightlessness: An Interview with Aleksey Leonov," *Soviet Military Review*, No. 6, June, 1981.

Barna, Rebecca, "Mechanical Metaphors For the Mind," *Datamation*, November 1, 1984.

Belew, Leland F., and Ernst Stuhlinger, *Skylab: A Guidebook*, NASA, Washington, D.C., 1973.

Beregovoy, G. T., L. S. Khachatur'yants, et al., *Activity of the Cosmonaut in Flight and Enhancing Activity Effectiveness*, Moscow, 1980.

Beregovoy, G. T., "Concerning the Question of the Role of the Human Factor in Space Flight," *Meditsinskaya Gazeta*, No. 1, January, 1983.

Beregovoy, G. T., I. V. Davydov, N. V. Drylova, I. B. Solovyeva, "Psychological Training—One of the Most Important Factors of Enhancing Safety of Space Flights," XXX Congress International Astronautical Federation, Munich, September 17-22, 1979.

Beregovoy, G. T., V. A. Popov, and V. S. Shustikov, "Research in Space Psychology," *Psikhologicheskiy Zhurnal*, Vol. 3, No. 4, Moscow, July–August, 1982.

Berezovoy, Anatoliy Nikolayevich, "211 Days in Orbit—His Memoirs," as told to V. Gorkoy and N. Konkov, *Aviatsiya I Kosmonavtika*, July, August, and September, 1983.

Berry, Adrian, *The Next Ten Thousand Years*, E. P. Dutton, New York, 1974.

Bluth, B. J., "The Truth about the Skylab 'Revolt,'" *L-5 News*, September, 1979, pages 12–13.

Boorstin, Daniel J., *The Discoverers: A History of Man's Search to Know His World and Himself*, Random House, New York, N.Y., 1983.

Boorstin, Daniel J., *The Americans: The Colonial Experience*, Random House, New York, N.Y., 1958.

Boston, Penelope, editor, *The Case for Mars*, American Astronautical Society, Univelt Incorporated, San Diego, Cal., 1984.

Brady, J., and Henry Emurian, "Experimental Studies of Small Groups in Programmed Environments," *Journal of Washington Academy of Sciences*, No. 73, 1983.

— References —

Brown, Dee, *The Gentle Tamers, Women in the Old West*, Putnam, New York, N.Y., 1958.

Burnazyan, Avadalik, "Man in Space," *Pravda*, Moscow, June 15, 1981, page 3.

Calder, Nigel, *Spaceships of the Mind*, Viking Press, New York, N.Y., 1978.

Cheston, Steven, and David Winter, editors, *Human Factors of Outer Space Production*, American Association for the Advancement of Science, Washington, D.C., 1980.

Clearwater, Yvonne, "A Human Place in Outer Space," *Psychology Today*, July, 1985, pages 34–43.

Cole, Dandridge, *Beyond Tomorrow*, Palmer Publications, Amherst, Wisconsin, 1965.

Collins, Michael, *Carrying the Fire: An Astronaut's Journeys*, Farrar, Straus, & Giroux, New York, N.Y., 1974.

Compton, W. David, and Charles D. Benson, *Living and Working in Space: A History of Skylab*, NASA SP-4208, Washington, D.C., 1983.

Connors, Mary M., Albert A. Harrison, and Faren R. Akins, *Living Aloft: Human Requirements for Extended Spaceflight*, NASA SP-483, Washington, D.C., 1985.

Connors, Mary M., Albert A. Harrison, and Faren R. Akins, "Psychology and the Second Space Frontier," in press, 1985.

Cooper, Henry S. F. Jr., *A House in Space*, Holt, Rinehart and Winston, New York, 1976.

Cunningham, R., D. Gennery, and E. Kan, "From Outer Space to Factory Floor," *Computers in Mechanical Engineering*, April, 1983.

Cunningham, Walter, *The All-American Boys*, Macmillan, New York, N.Y., 1977.

Derrick, William L., Thomas M. McCloy, Michael D. Matthews, Fran R. Wood, and Karen O. Dunivin, *Psychological, Sociological, and Habitability Issues of Long-Duration Space Missions*, Department of Behavioral Sciences and Leadership, USAF Academy, NASA contract number T-1082K, Johnson Space Center, Houston, Texas, 1985.

Dyson, Freeman, "Pilgrims, Saints, and Spacemen," *L-5 News*, May, 1979, pages 5–9.

Ehricke, Krafft A., "The Extraterrestrial Imperative," *Space World*, October, 1978, pages 4-22.

Ehricke, Krafft A., "The New Cosmos and Homo Extraterrestris," in *Our Extraterrestrial Heritage*, symposium proceedings, AIAA Los Angeles Section, Los Angeles, Cal., 1978.

Fisher, Anna Lee, and William F. Fisher, "Medical Implications of Space Flight," *TEM/Environmental Medical Emergencies*, March, 1980, pages 137-150.

Freitas, Robert A., Jr., and William P. Gilbreath, editors, *Advanced Automation for Space Missions*, NASA Conference Publication 2255, Washington, D.C., 1982.

Freitas, Robert A., Jr., and Patricia A. Carlson, *Computer Science: Key to a Space Program Renaissance*, 1981 NASA/ASEE Summer Study on the Use of Computer Science and Technology in NASA, The University of Maryland, Technical Report 1168, 1981.

Froelich, Walter, *Space Station: The Next Logical Step*, NASA EP-213, U.S. Government Printing Office, Washington, D.C., 1984.

— References —

Gatland, Kenneth, *The Illustrated Encyclopedia of Space Technology: A Comprehensive History of Space Exploration*, Harmony Books, Crown Publishers, New York, N.Y., 1981.

Gazenko, O. G, V. I. Myasnikov, K. K. Ioseliani, O. P. Kozerenko, and F. N. Uskov, "Important Problems of Space Psychology," Institute of Biomedical Problems, U.S.S.R. Ministry of Health, Moscow.

Gevarter, William B., *An Overview of Artificial Intelligence and Robotics: Artificial Intelligence, Applications*, NASA Technical Memorandum 85838, October, 1983.

Gevarter, William B., *An Overview of Computer Vision*, NASA and U.S. Department of Commerce, NBSIR 82-2582, September, 1982.

Gevarter, William B., *An Overview of Expert Systems*, NASA and U.S. Department of Commerce, NBSIR 82-2505, May, 1982.

Gevarter, William B., "Expert Systems: Limited but Powerful," *IEEE Spectrum*, August, 1983.

Goody, Richard, *Global Change: Impacts on Habitability*, A Report by the Executive Committee of a Workshop held at Woods Hole, Massachusetts, June 21–26, 1982, JPL D-95, 1982.

Gubarev, V., "His Heart Remains on Earth," *Pravda*, August 4, 1983, page 3; see also Lebedev, Valentin.

Gubarev, V., "The Elbrus's Orbits: An Interview with Cosmonaut V. A. Shatalov," *Pravda*, December 15, 1983, page 3.

Gubarev, V., "Two Months in Orbit: A Conversation with the Crew of Salyut-7," *Pravda*, August 27, 1983, page 3.

Hall, Stephen, Georg von Tiesenhausen, and Gary Johnson, *The Human Role in Space*, NASA Technical Memorandum TM-82482, Marshall Space Flight Center, Huntsville, Alabama, 1982.

Harrison, Albert, and Mary Connors, "Groups in Exotic Environments," in *Advances in Experimental Social Psychology*, edited by Leonard Berkowitz, Academic Press, San Francisco, 1984.

Harrison, Albert, and Mary Connors, "Psychological and Interpersonal Adaptation to Mars Mission," in press, 1985.

Harrison, Albert, "Case for Mars II: Report of the Psychology Task Force," in press.

Helmreich, Robert L., "Applying Psychology in Outer Space: Unfulfilled Promises Revisited," *American Psychologist*, No. 38, 1983.

Helmreich, Robert L., John A. Wilhelm, and Thomas E. Runge, "Psychological Considerations in Future Space Missions," in *Human Factors of Outer Space Productions*, edited by T. Stephen Cheston and David L. Winter, American Association for the Advancement of Science, Westview Press, Boulder, Colorado, 1980.

Heppenheimer, Thomas, *Colonies in Space*, Stackpole Publishers, Harrisburg, Pa., 1977.

Heppenheimer, Thomas, *Toward Distant Suns*, Stackpole Publishers, Harrisburg, Pa., 1979.

Houtchens, Bruce, *System for Management of Trauma and Emergency Surgery in Space*, NASA Grant NASW-3744, Houston, Texas, 1984.

Ivakhnov, A., "Protons, Do You Copy? Reportage from the Flight Control Center," *Izvestiya*, July 11, 1983, page 3.

— References —

Johnson, Nicholas L., *Handbook of Soviet Manned Space Flight*, American Astronautical Society, Univelt, Inc., San Diego, Cal., 1980.

Johnson, Philip C., and John A. Mason, editors, *Medical Operations and Life Sciences Activities on Space Station*, NASA Technical Memorandum 58248, Houston, Texas, 1982.

Johnson, Richard D., Daniel Bershader, and Larry Leifer, *Autonomy and the Human Element in Space*, Report of the 1983 NASA/ASEE Summer Faculty Workshop, Stanford, Cal., December 1, 1983.

Johnson, Richard D., "Space Habitats," in *Our Extraterrestrial Heritage*, symposium proceedings, AIAA Los Angeles Section, Los Angeles, Cal., 1978.

Jung, Carl G., *Memories, Dreams, Reflections*, Pantheon Books, New York, N.Y., 1973.

Kerwin, Joseph, "House Calls in Space: A Doctor-Astronaut's View," *Space World*, March, 1977, pages 26-29.

Konovalov, B., "Mistake Factor of Space Station Crew's Work Noted," *Izvestiya*, August 13, 1982, page 3.

Langereux, Pierre, "The 'Privileged' Space Cooperation between France and the Soviet Union," *La Recherche*, Paris, March, 1983.

Lebedev, Valentin, "Excerpts from In-flight Diary: His Heart Remains on Earth," with introduction by V. Gubarev, *Pravda*, August 15, 1983, page 3.

Leonov, A. A., "Prospects for the Conquest of Space and Psychology," in Petrov, 1979 (see below).

Leonov, A. A., and V. I. Lebedev, *Psychological Problems of Interplanetary Flight*, NASA Technical Translation, NASA TT F-16536, 1975.

Litsov, A. P., and I. F. Sarayev, "Questions of Ensuring Stable Human Working Capacity on a Long-Term Spaceflight," in Petrov, 1979 (see below).

Lunan, Duncan, *Man and the Planets*, Ashgrove Press, Bath, U.K., 1983.

Maguire, Bassett, Jr., "Ecological Problems and Extended Life Support on the Martian Surface," Case for Mars conference, Boulder, Colorado, 1981.

Malaurie, Jean, *The Last Kings of Thule*, translated by Adrienne Foulke, E. P. Dutton, New York, 1982.

Mark, Hans, "The Space Station—Mankind's Permanent Presence in Space," *Aviation, Space, and Environmental Medicine*, October, 1984, pages 948–956.

Martins, Gary R., "The Overselling of Expert Systems," *Datamation*, November 1, 1984.

Mashinskiy, Aleksandr, and Galina Nechitaylo, "The Birth of Plant Growing in Space," *Tekhnika-Molodezhi*, April, 1983, pages 2–7.

Mason, John A., and Philip C. Johnson, Jr., editors, *Space Station Medical Sciences Concepts*, NASA Technical Memorandum 58255, Houston, Texas, 1984.

Mason, Robert M., and John L. Carden, editors, *Controlled Ecological Life Support System*, NASA Conference Publication 2232, Washington, D.C., 1982.

McElroy, Michael, *Global Change: A Biogeochemical Perspective*, JPL Publication 83-51, July 15, 1983.

McKay, Christopher, editor, *The Case for Mars—II*, American Astronautical Society, Univelt Incorporated, San Diego, Cal., to appear.

— References —

Melenevskiy, I., "They Have Ordinary Dreams," *Trud*, Moscow, April 3, 1979, page 5.

Minsky, Marvin, "Telepresence," *OMNI*, June, 1980, pages 45-52.

Mullin, Charles S., Jr., and H. J. M. Connery, "Psychological Study at an Antarctic IGY Station," source unknown.

Muson, Howard, "The 'Right Stuff' Might Be Androgyny," *Psychology Today*, June, 1980, pages 14–18.

Myasnikov, V. I., "Mental Status and Work Capacity of Salyut-6 Station Crew Members," *Kosmicheskaya Biologiya I Aviakosmicheskaya Meditsina*, Vol. 17, No. 6, Moscow, November–December, 1983.

Myasnikov, V. I., and O. P. Kozerenko, "Prevention of Psychoemotional Disturbances during Long-Term Space Flights by Means of Psychological Support," *Kosmicheskaya Biologiya I Aviakosmicheskaya Meditsina*, No. 2, March–April, 1981.

Myasnikov, V. I., E. F. Panchenkova, F. N. Uskov, "Prospects of Using Radio and TV Communication Data in the Medical Supervision of Cosmonauts In-Flight," XXVth International Congress of Aviation and Space Medicine, Helsinki, Finland, September 4–9, 1977.

Nardini, J. E., R. S. Herrman, and J. E. Rasmussen, "Navy Psychiatric Assessment Program in the Antarctic," American Psychiatric Association 117th annual meeting, Chicago, Ill., May 8-12, 1961.

National Academy of Sciences, Space Science Board, *Human Factors in Long-Duration Spaceflight*, National Academy of Sciences, Washington, D.C., 1972.

National Aeronautics and Space Administration, *Earth Observing System: Science and Mission Requirements Working Group Report Appendix, Volume 1*, Technical Memorandum 86129, August, 1984.

National Aeronautics and Space Administration, *Earth Observing System: Status Report*, April 30, 1984.

National Research Council, Space Applications Board, *Practical Applications of a Space Station*, National Academic Press, Washington, D.C., 1984.

Nau, Dana, "Expert Computer Systems," *IEEE Spectrum*, February, 1983.

Nicogossian, Arnauld E., and James F. Parker, *Space Physiology and Medicine*, NASA SP-447, Washington, D.C., 1982.

Nikolayev, A., and M. Goryachev, "The Spaceflight . . . on Earth," *Kryl'ya Rodiny*, No. 3, 1979.

Novikov, M. A., "Principles and Methods of Studying Psychophysiological Compatibility," trans. by N. Timacheff, XIIth U.S./U.S.S.R. Joint Working Group Meeting on Space Biology and Medicine, Washington, D.C., November 9–22, 1978.

Novikov, M. A., and N. N. Gurovskiy, "Psychophysiological Screening—Status and Prospects," *Kosmicheskaya Biologiya I Aviakosmicheskaya Meditsina*, No. 2, March–April, 1981.

Novikov, N., "An Extended Expedition [Advantages of Long-Duration Space Flights]," in *Sovyetskiy Voin*, August, 1981, pages 28–29.

Oberg, Alcestis R., *Spacefarers of the '80s and '90s: The Next Thousand People in Space*, Columbia University Press, New York, N.Y., 1985.

Oberg, James E., *Red Star in Orbit*, Random House, New York, N.Y., 1981.

Oberg, James E., *Mission to Mars*, Stackpole Publishers, Harrisburg, Pa., 1982.

— References —

Oberg, James E., *The New Race for Space*, Stackpole Publishers, Harrisburg, Pa., 1984.

Office of Technology Assessment, *Civilian Space Stations and the U.S. Future in Space*, U.S. Congress, Technical Memorandum OTA-STI-242, Washington, D.C., 1984.

Office of Technology Assessment, *Salyut: Soviet Steps toward Permanent Human Presence in Space*, U.S. Congress, Technical Memorandum OTA-TM-STI-14, Washington, D.C., 1983.

Oleson, Mel, et al. [Boeing Aerospace Company], *Regenerative Life Support Research / Controlled Ecological Life Support System Program Planning Support (Transportation Analysis)*, NASA contract NAS2-11148, Ames Research Center, California, 1982.

Pankov, A., "A. N. Berezovoy Comments on Salyut-7 Crew's Work, Psychological Adaptation," *Leninskoye Znamya*, No. 1, January, 1983.

Parker, James F., Jr., and Vita R. West, managing editors, *Bioastronautics Data Book*, NASA SP-3006, NASA Scientific and Technical Information Office, Washington, D.C., 1973.

Pavlov, V. L., "Certain Aspects of the Interaction of Members of the Space Crew in Emergency Situations," in Petrov, 1979 (see below).

Pavlova, Marina, "Interview with Cosmonaut Svetlana Yevgen'yevna Savitskaya," *Nedelya*, no. 2, January, 1983.

Petrov, B. N., B. F. Lomov, and N. D. Samsonov, *Psikhologicheskiye Problemy Kosmicheskikh Polyotov (Psychological Problems of Space Flight)*, Nauka Press, Moscow, 1979.

Pishchik, V., "Commentary on Salyut-7 Crew's Medical Experiments," *Meditsinskaya Gazeta*, No. 41, May 18, 1984.

Pishchik, V., "Innovations in Cosmonaut Medical Monitoring, Physical Conditioning," *Meditsinskaya Gazeta*, April 11, 1984.

Pogue, William R., "The Eyes Have It," speech prepared for delivery to the Canadian Remote Sensing Society, Edmonton, Canada, Sept. 23, 1975.

Pogue, William R., *How Do You Go to the Bathroom in Space? and Other Questions*, Baen Books, New York, N.Y., 1985.

Pokrovskiy, A., "A Visit to the Doctors: A Report from Flight Control Center," *Pravda*, April, 1984.

Popovich, P. R., and Yu. I. Artyukhin, "Certain Characteristics of Crew Activity under Space Flight Conditions," in Petrov, 1979 (see above).

Quattrone, P. D., "Extended Mission Life Support Systems," paper AAS 81-237, in Boston, Penelope, editor, *The Case for Mars*.

Rasmussen, John E., "Group Behavior in Isolation—Antarctica," XIV International Congress of Applied Psychology, Copenhagen, Denmark, Aug. 19, 1961.

Remek, Vladimir, "Communication Problems of International Crews," XXX Congress International Astronautical Federation—Munich, Sept. 17–22, 1979.

Research Directorate of the National Defense University, *Climate Change to the Year 2000: A Survey of Expert Opinion*, National Defense University, Fort McNair, Washington, D.C., February, 1978.

Rice, Berkeley, "Space Lab Encounters," *Psychology Today*, June, 1983.

— References —

Robinson, George S., *Living in Outer Space*, Public Affairs Press, Washington, D.C., 1975.

Rohrer, John H., "Human Adjustment to Antarctic Isolation," Georgetown University School of Medicine.

Rose, Frank, *Into the Heart of the Mind: An American Quest for Artificial Intelligence*, Harper & Row, New York, 1984.

Rosenfeld, Anne H., "How Do the Soviets Spell Relief?", *Psychology Today*, July 1985, page 39.

Ross, Helen, "Dexterity Is Just a Fumble in Space," *New Scientist*, August 23, 1984, pages 16–17.

Ryumin, Valeriy, "Memoirs: Six Months above the Planet," trans. by Henry Gris, unpublished.

Ryumin, Valeriy, "Memoirs: Into Space—The Third Time," unpublished.

Santy, Patricia, "The Journey Out and In: Psychiatry and Space Exploration," *American Journal of Psychiatry*, May, 1983, pages 519–527.

Schiffer, Robert A., *Guidelines for the Air-Sea Interaction Special Study: An Element of the NASA Climate Research Program*, JPL/SIO Workshop Report, JPL Publication 80–8, Feb. 15, 1980.

Schlissel, Lillian, *Women's Diaries of the Westward Journey*, Schocken Books, New York, N.Y., 1982.

Seitz, William W., "Health Maintenance in Faraway Places," AIAA Symposium, Houston, Texas, 1983.

Sevastyanov, V. I., "The Appearance of Certain Psychophysiological Characteristics of Man under Conditions of Space Flight," in Petrov, 1979 (see above).

Shatalov, Vladimir, "The Psychological Problems of Long-Duration Flights," *Air and Cosmos*, September, 1981.

Simpson, Theodore R., editor, *The Space Station: An Idea Whose Time Has Come*, IEEE Press, New York, N.Y., 1985.

Staehle, Robert L., *Technologies for Space Station Autonomy*, NASA Jet Propulsion Laboratory, JPL Publication 84–85, Pasadena, Cal., 1984.

Stine, Harry, *The Space Enterprise*, Ace Books, Grosset & Dunlap, New York, N.Y., 1980.

Stratton, Joanna L., *Pioneer Women: Voices from the Kansas Frontier*, Touchstone Books, Simon and Schuster, New York, N.Y., 1981.

Stuster, Jack W., *Space Station Habitability Recommendations Based on a Systematic Comparative Analysis of Analogous Conditions*, Anacapa Sciences, Inc., for NASA, Moffett Field, California, 1984.

Stuster, Jack W., *Summaries of Alternative Analogues to a NASA Space Station and Description of Evaluation Methodology*, Anacapa Sciences, Inc., NASA Technical Report 553-2, 1984.

Sudakov, V., "Daily Life in Orbit: An Interview with Alexander Ivanchenkov," *Soviet Union*, No. 6, Moscow, June, 1984.

Tairbekov, M. G., and G. P. Parfenov, "Biological Research in Space," *Kosmicheskaya Biologiya I Aviakosmicheskaya Meditsina*, No. 2, March–April, 1981.

Tschopp, Alexander, and Augusto Cogoli, "Low Gravity Lowers Immunity to Disease," *New Scientist*, August 23, 1984, page 36.

— References —

Vasilyev, V., "Five-Month Experiment in Isolation Habitat with Biological Life Support," *Trud*, April 10, 1984, page 4.

von Tiesenhausen, Georg, *An Approach toward Function Allocation between Humans and Machines in Space Station Activities*, NASA TM-82510, Huntsville, Alabama, 1982.

von Tiesenhausen, Georg, *Management and Control of Self-Replicating Systems: A Systems Model*, NASA Technical Memorandum TM-82460, February, 1982.

Wolbers, Harry L., and Stephen B. Hall, study managers, *THURIS: The Human Role in Space*, NAS 8-35611, McDonnell Douglas Astronautics Company, Huntington Beach, Cal., 1984.

Wolfe, Tom, *The Right Stuff*, Farrar, Straus & Giroux, New York, N.Y., 1979.

Yankulin, V., "Life in Orbit—An Interview with Oleg Gazenko, Director of the Institute of Biomedical Sciences," *Nauka I Zhizn*, No. 3, 1979.

Zubkov, V., "Interview with the Cosmonauts Who Made the 211-Day Orbital Mission," *Sotsialisticheskaya Industriya*, No. 295, December, 1982.

index